DNA, GENES AND
GENETIC ENGINEERING

ABOUT THE AUTHORS

Gordon R. Carter, D.V.M., M.S., D.V.Sc.

Dr. Carter is a Professor Emeritus of the Department of Biomedical Sciences and Pathobiology at the Virginia-Maryland Regional College of Veterinary Medicine, Virginia Polytechnic Institute and State University in Blacksburg, Virginia. Professor Carter, whose principal interest is pathogenic microbiology, has published many scientific papers and is the senior author of more than a dozen textbooks. He has served on the faculties of the University of Toronto and Michigan State University, and acted as a consultant on infectious diseases of animals for several international agencies in various countries of Asia, Africa, and South America.

Stephen M. Boyle, B.A., Ms., Ph.D.

Dr. Boyle is a Professor of Microbiology in the Department of Biomedical Sciences and Pathobiology at the Virginia-Maryland Regional College of Veterinary Medicine, Virginia Polytechnic Institute and State University. Professor Boyle, whose principal research interests are microbial virulence mechanisms and the application of recombinant DNA technology to vaccine development, has authored numerous papers in the areas of microbial gene regulation. He has served on the Faculty of Medicine at Memorial University of Newfoundland, and is Vice-President of Veterinary Technologies Corporation, a biotech company located at the Virginia Tech Corporation Research Center and dedicated to the development and marketing of veterinary vaccines.

All You Need to Know About

DNA, GENES AND GENETIC ENGINEERING

A Concise, Comprehensive Outline

By

GORDON R. CARTER

STEPHEN M. BOYLE

CHARLES C THOMAS • PUBLISHER, LTD.
Springfield • Illinois • U.S.A.

Published and Distributed Throughout the World by

CHARLES C THOMAS • PUBLISHER, LTD.
2600 South First Street
Springfield, Illinois 62794-9265

© *1998 by* CHARLES C THOMAS • PUBLISHER, LTD.
ISBN 0-398-06870-4

Library of Congress Catalog Card Number: 98-3519

Printed in the United States of America
CR-R-3

Library of Congress Cataloging-in-Publication Data

Carter, G. R. (Gordon R.)
    All you need to know about DNA, genes, and genetic engi-
neering: a concise. comprehensive outline / by Gordon R. Carter,
Stephen M. Boyle.
        p.   cm.
    Includes bibliographical references and index.
    ISBN 0-398-06870-4 (paper).
    1. Molecular genteics--Outlines, syllabi, etc. 2. Genetic engi-
neering--Outlines, syllabi, etc. 3. Medical genetics--Outlines, syllabi,
etc.   I. Boyle, Stephen M. II. Title.
    QH442 C37 1998
    572. 8--dc21
                                                                98-3519
                                                                    CIP

Any subject may be made repulsive
by presenting it bristling with difficulties

Sylvanus P. Thompson
Author of *Calculus Made Easy*

PREFACE

Knowledge of molecular genetics has increased dramatically since the epochal discovery of the structure of DNA by J.D. Watson and F. Crick in 1953. By now we are well into a Genetic Revolution which has been compared for ultimate significance to the Industrial Revolution. The applications of the new genetic technologies are manifold and future developments would seem virtually unlimited.

Among the benefits thus far are new vaccines, improved plants and animals, better and more economic ways of producing hormones, drugs, and many commercially useful proteins. The causes of many genetic disorders have been elucidated and tests for the genes that may predispose individuals to such maladies have and are being developed. It is clear that the many applications of **Genetic Engineering**, some of which are already controversial, will profoundly impinge all of our lives.

The purpose of this outline is to provide, for students and the general reader, the basic ideas and facts relating to **DNA, Genes,** and **Genetic Engineering**, in a simple, uncluttered, understandable form. The great mass of technical information on the subject has been stripped to its essentials and presented in a summary fashion. The many practical applications have received particular emphasis. In short, this outline provides a painless way to obtain a comprehensive grasp of this very important subject.

There are a number of books on the subject for the laity, but some are too technical, others are too detailed, and none we have come across is comprehensive enough from a practical standpoint.

Persons with only a smattering of chemistry and biology should have no difficulty in understanding the ideas and following the procedures. The **Glossary**, which is particularly comprehensive, should be especially helpful.

We wish to express our thanks to Gerald A. Baber and Thomas A. Capone of the Biomedical Media (Virginia-Maryland Regional

College of Veterinary Medicine-VMRCVM) for preparation of many of the illustrations, and to Dr. Darla J. Wise for her helpful suggestions and careful review of the manuscript. We are indebted to the staff of Information Processing (VMRCVM) for preparing the final manuscript.

GORDON R. CARTER
STEPHEN M. BOYLE

CONTENTS

DNA, GENES AND
GENETIC ENGINEERING

Chapter 1

NUCLEIC ACIDS, CHROMOSOMES, GENES, AND MITOCHONDRIA*

Information responsible for the characteristics of all organisms is stored in the long, slender molecule of deoxyribonucleic acid (DNA). This long molecule is made up of regions called genes that code for proteins which are made up of long chains of amino acids. The DNA molecule with its genes and some other supporting material constitute the structure called the chromosome. Depending on the cell type, there can be more than one chromosome, e.g., in humans there are 46 per cell.

DNA (DEOXYRIBONUCLEIC ACID)

DNA consists of two long strands of linked nucleotides (bases). The shape of the entire molecule is that of a double helix. The sugar (deoxyribose) and the phosphate form the backbone of DNA, while the bases lie flat in the center of the molecule like the steps of a staircase. The two strands are held together by weak hydrogen bonds which allow certain base pairings: adenine/thymine (A/T) and guanine/cytosine (G/C) (see Figures 1.1, 1.2, and 1.3).

---

*Terms not explained in the text are defined at the end of chapters or in the Glossary.

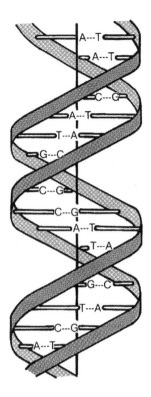

Figure 1.1 The DNA molecule consists of a pair of sugar-phosphate backbones wound around each other in the form of a double helix. The nucleotides, or bases protrude inward like the rungs of a ladder. The order in which the nucleotides occur in a single strand of DNA varies from one DNA molecule to another, however, they follow strict rules of pairing as is shown: cytosine (C) on one strand pairs with guanine (G) on the opposite strand, and adenine (A) pairs with thymine (T). C always pairs with G, and A always pairs with T. U.S. Congress, Office of Technology Assessment, *Technologies for Detecting Heritable Mutations in Human Beings,* OTA-H-298 (Washington, DC: U.S. Government Printing Office, September 1986).

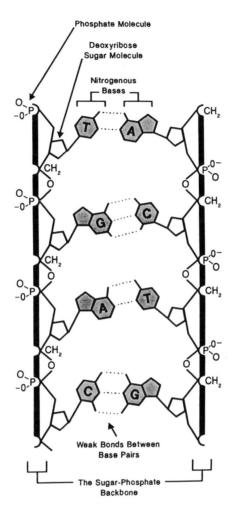

Figure 1.2. The four nitrogenous bases, adenine (A), guanine (G), cytosine (C), and thymine (T), form the four letters in the alphabet of the genetic code. The pairing of the four bases is A with T and G with C. The sequence of the bases along the sugar-phosphate backbone encodes the genetic information. Reprinted from U.S. Congress, Office of Technology Assessment. *Mapping Our Genes—The Genome Project: How Big? How Fast?* OTA-BA-373 (Washington, D.C.: U.S. Government Printing Office. April 1988).

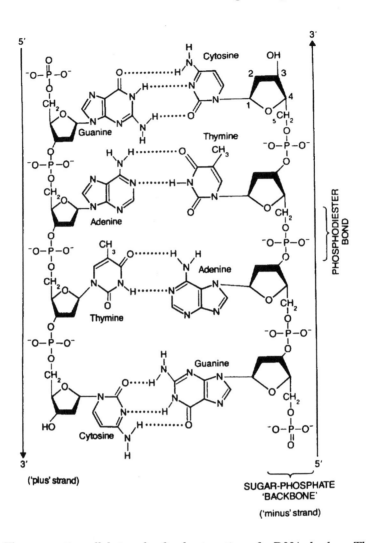

Figure 1.3. The two antiparallel strands of a short section of a DNA duplex. The strands actually twist around each other in a double helix. The dots linking the central base-pairs represent hydrogen bonds. Primed numbers (5', 3') indicate the carbon atom numbers on the deoxyribose moiety involved in the phosphodiester bonding of each chain. These carbon atoms are numbered in the deoxyribose moiety attached to the top cytosine. The left-hand chain runs from 5' to 3' top to bottom, the right-hand chain runs from 5' to 3' bottom to top. From *Dictionary of Biology*, 9th ed. by M. Thain and M. Hickman. Penguin Books, New York, N.Y. 1994.

Unlike many biological molecules, double-stranded DNA is particularly resistant to deterioration over time. Single-stranded DNA is less resistant to damage. Some viruses have single-stranded DNA or RNA while others have double-stranded DNA or RNA. Most plant viruses have RNA.

Sources of DNA

All cells but mature mammalian red blood cells possess DNA. Thus the sources of DNA are many and which ones are used will depend on the use or application intended. These sources for humans include: various tissue samples, skin, hair root, blood (cells other than erythrocytes), semen (spermatozoa), cells from saliva collected by swabbing the mouth (cheeks and gums) with a soft brush, and tissue samples, including blood, from the fetus.

Isolation of DNA

In order to work with DNA, it is first necessary to isolate it. DNA is a long, stringy molecule that can be readily separated from most other cellular constituents. One procedure involves the following steps:
- Treat the sample with detergent to break open the cells; the ease with which this can be done depends on the cell type.
- Salt may be added to precipitate the proteins which are then removed by centrifugation.
- The remaining proteins can be broken down by enzymes (proteases).
- Phenol may be added to facilitate removal of contaminants; the DNA solution readily separates from the phenol phase containing the contaminants.
- Alcohol is added carefully and results in precipitation of the DNA at the boundary of the alcohol and DNA solution.
- The sticky DNA, which resembles nasal mucus, is collected by winding it onto a glass rod (see Figure 1.4).

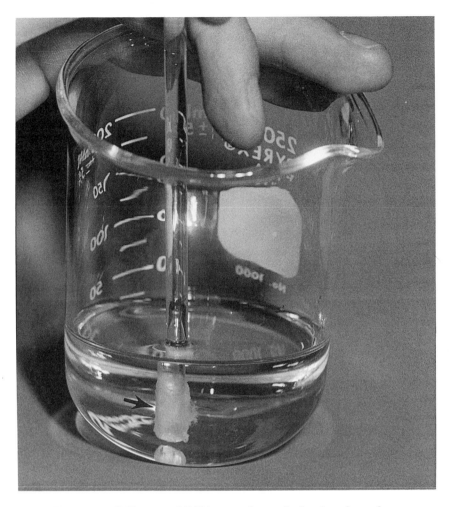

Figure 1.4 Collection of DNA on a glass rod after its release from
cells treated with a detergent.

RNA (RIBONUCLEIC ACID)

RNA molecules differ from DNA in usually being single-stranded.
RNA is structurally similar to DNA in having three subunits:
  • a phosphate group
  • a 5'-carbon (pentose) sugar ribose
  • a nucleotide: either adenine(A), guanine (G), cytosine (C), or
    uracil (U)
In RNA, A pairs with U, and G pairs with C.

There are several kinds of RNA that serve various functions, including "translating" selected gene sequences coded in the cell's DNA. It has recently been found that certain forms of RNA can act as enzymes and thus carry out their own replication. They can both carry genetic messages and act as an enzyme (ribozyme).

CHROMOSOMES

A chromosome is a microscopic, thread-like, self-replicating body composed of chromatin. In eukaryotes (organisms with a nucleus), the chromosomes are located in the nucleus. Each chromosome contains a long, linear DNA molecule which contains genes.

Each animal species has a set number of chromosomes. Humans have 46 chromosomes, 22 matching pairs (autosomes) plus two sex chromosomes. The female has two X chromosomes, the male has an X and a Y chromosome. The sex of a human is determined by the X and Y chromosomes. Human chromosomes are numbered in order of decreasing size (1 to 22) except that 22 is larger than 21. Chromosome 1 has about 300 million base pairs while 21 has about 50 million.

Chromosomes exist as homologous pairs (diploid) in all eukaryotic cells except sex cells (egg and sperm); the latter have one chromosome of each type (haploid). Prokaryotic cells have, with few exceptions, only one chromosome and are thus haploid.

There is great variation in the number of chromosomes in species of eukaryotes. The haploid number for the fruit fly is 4, corn 10, yeast 16, nematode (a small round worm) 12, mouse 20, and chicken 39.

The fine thread (molecule) of DNA within each chromosome is about 100,000 times as long as its chromosome. The chromosomes of a human cell contain approximately three billion nucleotide base pairs.

Most bacteria have one chromosome consisting of a single, circular loop, which is folded and attached to the plasma membrane. The size of the genome (all of the genetic information of an organism) of *Escherichia coli* (a common intestinal bacterium) is about four million base pairs. Mycoplasmas, some species of which cause disease, are the smallest bacteria. Their genomes are about one-fifth the size of that of *E. coli* and are about 800,000 base pairs in length.

The single-celled microorganisms called archaebacteria or archaea, are no longer considered to be bacteria. They have been found to be distinct from eukaryotes (organisms whose cells have a nucleus) and prokaryotes (bacteria). The genome of one species *Methanococcus jannaschii*, recovered from hot springs, has been sequenced and found to resemble eukaryotes more closely than prokaryotes. Fifty-six percent of its 1,738 genes are unlike anything observed in eukaryotes or prokaryotes. The archaea are a greatly varied group that can live in very forbidding environments.

The DNA of chimpanzees and humans differs by only about 1 percent and there is much similarity with the DNA of other mammals.

GENES

A gene is a hereditary determinant of a trait or characteristic transmitted via the chromosome. It is a segment of DNA that codes for one polypeptide. Genes vary considerably in size: small ones may consist of about 1000 nucleotides; large ones may have as many as two million nucleotides. They regulate where, when and how much protein is produced.

About 5 percent of human genomic DNA is transcribed; 95 percent or more is silent, although some sequences are responsible for such cellular functions as meiosis, replication, etc.

An average gene (= 1 kilobase—one thousand bases) encodes a protein of approximately 35,000 Daltons (a unit equal to the mass of the hydrogen atom). The human genome has about 100,000 genes, *E.coli* about 2,000, the small virus of HIV (human immunodeficiency virus) has nine, the bacterial virus lambda about 50, and the large virus that causes cowpox about 200.

When genes inherited from one parent are more active than genes inherited from the other parent, they are referred to as imprinted genes. The molecular mechanisms involved are still being investigated. DNA methylation is sometimes involved; the more methylation, the less expression. Methylated or imprinted genes ordinarily affect the early development and growth of the embryo. Thus far they have only been found in mammals and flowering plants.

MITOCHONDRIA

These are minute, cylindrical, cytoplasmic organelles of eukaryotic cells. They are involved in aerobic respiration and are the source of most ATP (adenosine triphosphate) in cells. That they are descendants of bacteria was recognized as early as 1918 and is suggested by their prokaryotic type ribosomes. Mitochondria (singular: mitochondrion) vary in number in cells from one to several thousand and have a circular DNA molecule (about 16,000 base pairs in mammals). At fertilization the female's mitochondrial DNA is passed to succeeding cells; because of the absence of cytoplasm in spermatozoa, the male's is not passed on.

The fact that children have the same mitochondrial DNA as their mother has been used to confirm the maternity of children, e.g., the children of the "disappeared" in Argentina. A disease called Leber's optic neuropathy was the first genetic disorder to be traced to a defect (a single mutation) in mitochondrial DNA. This rare disorder is characterized by loss of vision as cells of the optic nerve gradually die.

As research in mitochondrial genetics increases, it seems likely that this DNA will be implicated in more genetic disorders.

Chapter 2

CENTRAL DOGMA, THE GENETIC CODE, AND PROTEIN SYNTHESIS

CENTRAL DOGMA

This is the concept that hereditary information is encoded in the DNA sequence and, with the exception of RNA viruses and prions, this information flows from DNA to RNA resulting in the production of protein. Steps in the process are:
- replication: copying DNA to make more DNA
- transcription: copying DNA to make messenger RNA
- translation: deciphering or decoding messenger RNA to make protein

RNA viruses transcribe single-stranded DNA from RNA templates by the enzyme reverse transcriptase.

The genetic information in DNA can be altered by mutation, i.e., changes in the composition of DNA. Mutations are an important element of heritable variation upon which natural selection operates. This means that if a mutation occurs which improves the fitness of an organism, the organism is likely to have selective advantage in competition with nonmutated organisms.

REPLICATION

In the process of replication, the two strands of DNA separate to make a Y-shaped replication fork (see Figure 2.1). Each strand can act as a template for formation of a new DNA strand. Free nucleotides

13

align along each DNA strand according to the complementary base-pairing rule (see Figure 1.2) and they are joined by the action of DNA polymerase to form a new DNA strand. Thus each new molecule is half newly synthesized and halfconserved (semiconservative replication). The enzyme DNA polymerase mediates formation of the new DNA using information from the old DNA (see Figure 2.2).

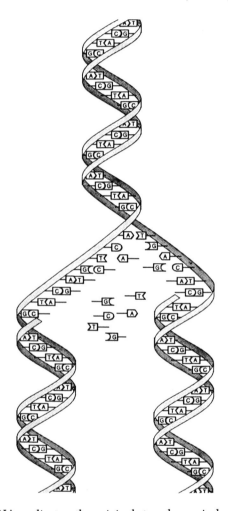

Figure 2.1. When DNA replicates, the original strands unwind and serve as template for the building of new, complementary strands. The daughter molecules are exact copies of the parent, each daughter having one of the parents strands. Reprinted from U.S. Congress, Office of Technology Assessment. *Mapping Our Genes–The Genome Project: How Big? How Fast?* OTA-BA-373 (Washington, D.C.: U.S. Government Printing Office. April 1988).

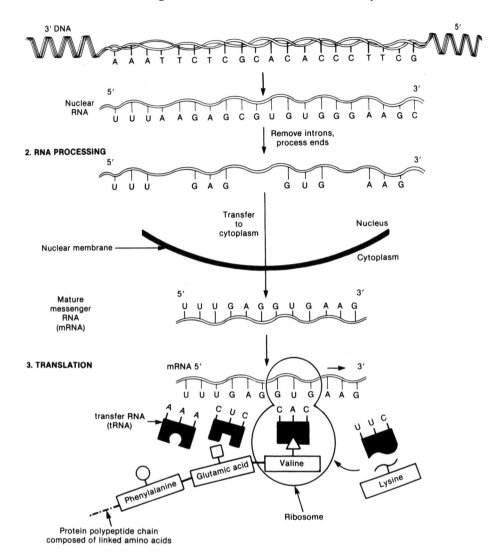

Figure 2.2. Protein Synthesis (in higher organisms): In the instruction process (i.e., coupled transcription and translation), a segment of DNA is transcribed into an RNA molecule. The introns are then snipped out, and the mature messenger RNA (mRNA) molecule is moved out of the cell's nucleus to a ribosome. In the translation process, transfer RNA (tRNA) molecules help translate the genetic code in the mRNA into amino acids, which are linked together to form a peptide or protein. Reprinted from: U.S. Congress, Office of Technology Assessment, *Technologies for Detecting Heritable Mutations in Human Beings*, OTA-H-298 (Washington, D.C.: U.S. Government Printing Office, September, 1986).

In cell division replication involves all of the chromosomes. Many hundreds of localized acts of DNA replication take place during the same time period; it is a very complicated process.

The active zones of DNA duplication are known as replicons. They behave as autonomous units during DNA replication and consist of 50,000 to 3,000,000 base pairs. The bacterial chromosome functions as a single replicon while eukaryotic chromosomes have many in series.

TRANSCRIPTION

The two strands of the DNA molecule separate over a certain region along their length (see Figure 2.2). The unpaired bases that make up the broken staircase attract free-floating nucleotides in the cell. The nucleotides are those of RNA, and a matching strand of messenger RNA (mRNA) is produced by the action of RNA polymerase. The order of the nucleotides is determined by the complementary order of the nucleotides of the DNA strand (template).

Excision removes some intervening sequences (introns) that will not be translated into protein (see Figure 2.3). The remaining shorter sequences joined together are called exons. They determine the sequence of amino acids in proteins. Bacteria lack the intronic sequences found in eukaryotic genes and also the splicing mechanism that removes the intronic sequences during processing in mRNA.

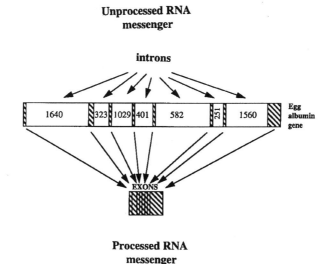

Figure 2.3. Removal of introns followed by splicing of exons. The egg albumin gene consists of seven introns and eight exons. The seven introns are eliminated during the formation of messenger RNA. The RNA message contains only the sequence of bases that specifies the amino acid sequence in the protein egg albumin.

A medium-size gene, albumin, has about 25,000 nucleotides; the mRNA consists of 2,100 nucleotides and the number of introns is 14; a large portion of the gene is "silent."

TRANSLATION

Translation is the process of converting the information in mRNA into protein (see Figure 2.2). The amino acid sequences in proteins are determined by the base sequences in the mRNA template on the ribosome (see below). The coding units in mRNA are codons; each codon contains three bases (triplet code).

Each codon of the mRNA corresponds with an anticodon on the transfer RNA (tRNA); this match-up or joining (complementary base pairing) is accomplished by a special enzyme, aminoacyl-tRNA synthetase. There is one synthetase for each different amino acid; each synthetase is responsible for joining the correct amino acid to its corresponding tRNA.

The tRNA serves as an adaptor which controls the match-up of codons (in the mRNA) and amino acids. The order of the codons determines the order of the amino acids in the new protein chain.

The genes of almost all organisms "communicate" with ribosomes and transfer RNA in the same genetic language. Exceptions are the mitochondria that have their own very small chromosome, and "prions" (infectious protein).

Birds have about half the amount of DNA as mammals and reptiles; large portions of introns are missing in birds.

Ribosomes

These very small particles, found in both prokaryotic and eukaryotic cells, are made up of special RNA molecules (ribosomal RNA/rRNA) and about 50 specific proteins (ribosomal proteins). Proteins are made in these particles or ball-like structures. Each ribosome consists of two "balls" or subunits; in prokaryotes, the 70S ribosome consists of a small one, 30S, and a larger one 50S (S = Svedberg unit, an expression of the sedimentation rate during ultra-high-speed centrifugation). The 80S ribosome of eukaryotes consists of subunits 40S and 60S. There usually are many ribosomes in a cell; prokaryotic cells may have tens of thousands. In eukarotes, they are attached to the endoplasmic reticulum but also are free in the cell.

Several antibiotics act by inhibiting protein synthesis on the ribosomes, and because of the differences between prokaryotic and eukaryotic ribosomes, the prokaryotic cells are killed while the eukaryotic cells are unaffected.

Ribosomes mediate the translation of mRNA into protein (see Figure 2.2). The set of relationships between the nucleotide base-pair triplets of a messenger RNA molecule and the 20 amino acids that are the building blocks of protein constitute the genetic code. With few exceptions, referred to above, the genetic code is universal for all living things.

There is a precise set of rules for translating codons into amino acids and thus proteins (see Table 2.1 The Genetic Code). The code is "commaless," i.e., nothing separates one codon from the next. The code is degenerate or redundant in that different codons may code for the same amino acid, e.g., CGA,CGG,CGC, and CGU all code for arginine. The average protein is composed of about 600 amino acids.

Table 2.1

THE GENETIC CODE

(In Which Messenger RNA Codons are Matched wtih Amino Acids)

First Base (5' end)	U	C	Second Base	A	G	Third Base (3' end)
U	UUU Phe	UCU Ser		UAU Tyr	UGU Cys	U
	UUC Phe	UCC Ser		UAC Tyr	UGC Cys	C
	UUA Leu	UCA Ser		UAA Stop	UGA Stop	A
	UUG Leu	UCG Ser		UAG Stop	UGG Trp	G
C	CUU Leu	CCU Pro		CAU His	CGU Arg	U
	CUC Leu	CCC Pro		CAC His	CGC Arg	C
	CUA Leu	CCA Pro		CAA Gln	CGA Arg	A
	CUG Leu	CCG Pro		CAG Gln	CGG Arg	G
A	AUU Ile	ACU Thr		AAU Asn	AGU Ser	U
	AUC Ile	ACC Thr		AAC Asn	AGC Ser	C
	AUA Ile	ACA Thr		AAA Lys	AGA Arg	A
	AUG Met	ACG Thr		AAG Lys	AGG Arg	G
G	GUU Val	GCU Ala		GUA Asp	GGU Gly	U
	GUC Val	GCC Ala		GAC Asp	GGC Gly	C
	GUA Val	GCA Ala		GAA Glu	GGA Gly	A
	GUG Val	GCG Ala		GAG Glu	GGG Gly	G

Table 2.1 The Genetic Code

There are three "stop" or chain-terminating codons: UAA,UAG, and UGA.

Ribosomal RNA exhibits more homology among widely dissimilar organisms than does DNA. This is because the nucleotide sequences of the rRNA genes are highly conserved, i.e., they have changed more slowly than the DNA. Ribosomal RNA similarity values (based on sequences of rRNA) have been used in determining evolutionary relatedness of many types of bacteria, plants, and animals.

Chapter 3

CONTROL OF GENE EXPRESSION

GENERAL

The details of this complex subject are mainly of interest to the molecular biologist. Some very general considerations are provided below. Gene expression in general is discussed in Chapter 2 under Transcription.

A major problem in understanding an organism's development is how the pattern of the differential use of genes in various tissues occurs. Determining how a gene is expressed involves the study of the RNA transcript and how it is translated during the life cycle of the organism. Most transcripts are copies of the gene and in higher organisms the transcripts are processed prior to translation.

A set of proteins (DNA-binding proteins) must act on the regulatory region of a gene before it can be transcribed into an RNA transcript; the regulatory region or domain is often located next to the gene. Some of the sets of proteins respond to hormonal and enviromental signals while others are specific for tissues.

The domain that locates where transcription begins in eukaryotes is the same from one gene to another. It includes a series of alternating A-T and T-A base pairs; this domain is referred to as the TATA box.

The mechanism that initiates a gene's transcription is very complex and varies from gene to gene. Cells of a variety of tissues make regulatory proteins that can activate a gene specific for a particular organ, but only the cells of that organ have all the proteins required to signal initiation of transcription involving that gene.

The control of gene expression varies considerably between prokaryotes and eukaryotes. Readers whose interests are mainly practical may wish to skip what follows.

21

IN PROKARYOTES

The cell uses various mechanisms to control its own metabolism. The gene for a particular enzyme must be expressed or turned on and then turned off when not required. A number of gene control systems have been shown in bacteria. Three important ones are:

1. The inducible operon
2. The repressible operon
3. The CAP-cAMP system

The term operon refers to groups of prokaryotic genes regulated together as a unit (see Figure 3.1) . They are all turned on or off at the same time (coordinate regulation) and their expression is necessary to carry out a specific function.

The *lac* operon of *E. coli* contains in order:

• A repressor gene *lacI*
• A promoter *lacP*
• An operator *lacO*
• 3 structural genes: *lacZ, lacY, lacA*

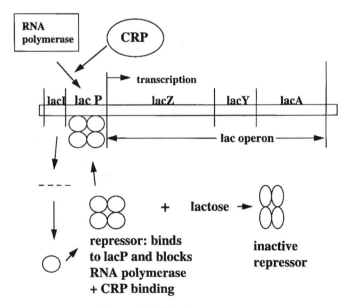

Figure 3.1 The Lac operon.

The enzyme ß-galactosidase is encoded by the *lac* operon. The *lac* operon promoter does not easily attract RNA polymerase which slows enzyme (i.e., ß-galactosidase) production. The CAP (catabolite activator protein) increases the affinity of RNA polymerase for the *lac* promoter. CAP alone is not very active, however, when cAMP (cyclic adenosine monophosphate) binds to it and the complex attaches to the *lac* promoter site, it increases the ability of RNA polymerase to bind and transcribe the *lac* gene. As lactose concentration increases, *lac* repressor molecules become more active and the cAMP levels increase. This increase means more CAP-cAMP complexes are available to keep the lac operon turned on.

For expression of a cloned gene (see Chapter 4) in a prokaryotic host cell a promoter region (for initiating transcription) and a terminator region (for cessation of transcription) must be in the correct locations.

Expression systems developed for protein synthesis in an industry setting are quite diverse. Each system involves the development of conditions that favor the production of a particular protein.

IN EUKARYOTES

The control of gene expression in eukaryotes is not well understood. Some important points are the following:
- DNA methylation, which can occur during replication, can cause DNA to be inactive.
- Unnecessary genes in specialized cells tend to be more methylated than active genes.
- Most genes in eukaryotic cells are not expressed (about 99 percent) and for this reason systems for activating transcription are very significant.
- As in bacteria, regulator genes have been found in eukaryotes that produce molecules that act as repressors or activators; promoter regions interact with these molecules.
- Much eukaryotic DNA is bound to histones; when DNA is thus bound, histones may serve as a means of regulation.

Chapter 4

RECOMBINANT DNA AND GENE CLONING

RECOMBINANT DNA TECHNOLOGY
(GENETIC ENGINEERING)

This refers to the techniques used in the construction, study, and use of recombinant DNA molecules. A recombinant DNA molecule is one with a novel DNA sequence produced *in vitro* ("in glass") by joining together pieces of DNA from different organisms.

Gene cloning is the technique of joining *in vitro* a gene to a vector, e.g., a plasmid, which is then transferred into a bacterium or other host so that the gene can be replicated and/or expressed.

STEPS IN THE GENE CLONING PROCESS

(Plasmid and Restriction Endonuclease are explained at the end of this section.)

- A restriction endonuclease (cuts DNA molecules only at a limited number of specific nucleotide sites) is used to "cut-up" the DNA molecule from, e.g., a human source (see Figure 4.1).
- A restriction endonuclease is also used to excise a piece of DNA from the body of the plasmid, other vector, or chromosome.
- When the DNA has been cut into pieces, a small segment, a gene or several genes in length, are separated based on size.
- The piece of human DNA and the plasmid DNA have "sticky ends" as a result of the treatment with endonuclease.
- The piece of human DNA is annealed to the plasmid ends and joined with DNA ligase.

- The recombinant plasmid is then used as the vector to carry the DNA into a host cell, usually a bacterium.
- The bacterium is then grown producing large numbers of identical chimeric bacteria which are called clones.

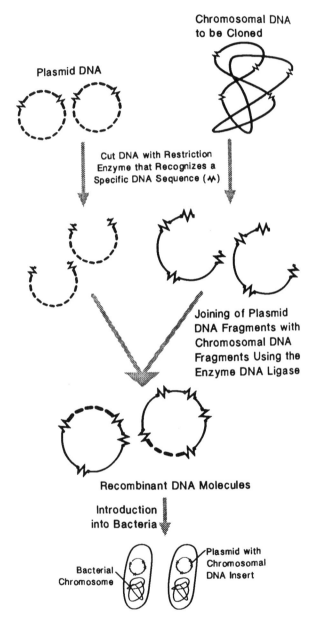

Figure 4.1. DNA cloning in plasmids. Reprinted from: U.S. Congress, Office of Technology Assessment, *Mapping Our Genes—The Genome Project: How Big, How Fast?* OTA-BA-373 (Washington, D.C.: U.S. Government Printing Office. April 1988).

OBTAINING A CLONE OF A SPECIFIC GENE

There are two approaches:

Direct Selection for the Desired Gene:

- The cloning experiment is designed so that the only clones are those containing the desired gene.
- Selection usually occurs at the plating-out stage, e.g., only recombinant bacteria carrying the antibiotic resistance gene borne in the vector can grow on a medium containing the antibiotic.
- If the recombinant bacteria carry a gene that codes for the enzyme that synthesizes tryptophan, the recombinant will grow on a medium without tryptophan; bacteria requiring tryptophan (auxotrophs) will not grow on the medium.

Identification of the Clone from a Gene Library:

There are many important genes that cannot be cloned by direct selection in bacteria, particularly genes from animals and plants. Identification of a clone from a genomic library is an alternative approach.

- A genomic library (see explanation below) is a set of recombinant clones that contain all or most of the DNA (genes) present in a particular organism.
- Genomic libraries are prepared by obtaining purified total cell DNA; making a partial restriction digest of the DNA; and inserting the fragments into a suitable vector.
- The bacteria are then grown producing large numbers of identical chimeric bacteria, each colony representing a clone with a unique segment of the genome.

SELECTING CELLS WITH THE PLASMID-CONTAINING GENE OF INTEREST

Using Antibiotic Resistance Genes:

A plasmid having genes encoding resistance to two antibiotics:
• One drug-resistant (e.g., ampicillin) gene on the plasmid has a restriction endonuclease cleavage site where DNA segments can be inserted; insertion of a DNA segment into this site will inactiate the drug-resistant gene (Insertional Inactivation, see Chapter 8).
• The second drug-resistance gene (tetracycline) is still active.
• The recombinant plasmid DNA molecules will confer resistance to bacteria with respect to only the second drug, tetracycline.
• The bacteria with the recombinant DNA will grow on the medium containing tetracycline but not on the medium containing ampicillin.
• The bacteria that grow in the presence of tetracycline are screened for the gene of interest (see below).

An Immunological Test for the Protein Encoded by the Cloned Gene of Interest:

This involves producing an antibody for the protein, labeling the antibody, and using it to detect the protein in recombinant bacteria.

Labeled protein A, which recognizes antibody, can also be used to determine whether or not the antibody has coated the recombinant bacteria.

IDENTIFICATION OF THE SPECIFIC GENE

The recombinant bacteria are tested for the presence of the specific gene by DNA hybridization (see below) with a labeled complementary DNA probe. If a radioactive label is used autoradiography is employed for detection. Nonradioactive labels may also be used (see Chapter 10).

The immunological procedure (an immunoblot) described above can also be used to identify a specific gene.

Ways of obtaining probes for detecting specific genes are discussed below.

OBTAINING A PROBE FOR A GENE

If the translation product, a protein, is known, its amino acid sequence can be determined and one can use the genetic code to predict the nucleotide sequence of the gene. Short oligonucleotides of predetermined sequence can be synthesized *in vitro* and used to determine the gene coding for the protein being studied. Because all but two amino acids are coded by at least two codons, the prediction will be approximate, but enough nucleotide homology may be shared to produce a clear hybridization result. DNA hybridization is explained below.

Two genes for the same protein from different organisms may be sufficiently homologous that a single-stranded probe prepared from the one gene will hybridize with the second gene; this is called heterologous probing. Although heterologous probes may not be entirely complementary, sufficient base-pairing can occur for a hybrid to be formed and recognized by autoradiography (see Glossary).

Various markers or labels other than a radioactive one are used to label heterologous probes. The label is either attached (e.g., a fluorescent dye) to or incorporated in (a radioactive compound or atom) the macromolecule, frequently DNA.

CLONING DNA SEGMENTS THAT ENCODE EUKARYOTIC PROTEINS

DNA segments containing genes that encode eukaryotic proteins are obtained as follows:
- Purified mRNA specific for the protein is used as a template for reverse transcriptase to synthesize a complementary DNA strand.
- This strand then acts as a template for DNA polymerase to produce a second complementary DNA strand.
- After enzymatic treatment this double stranded cDNA can be cloned into a vector.

GENETIC TRANSFORMATION OF PROKARYOTES

Transformation is the introduction of free DNA, e.g., a plasmid, into a living cell whereby it is either (1) integrated into the cell's genome or (2) not integrated but replicates; in both cases the DNA is often expressed. Some of the procedures used to transform bacterial cells follow:

- Treatment of bacteria with cold calcium chloride solution followed by short exposure to a high temperature. How this procedure facilitates transformation is not known.
- Electroporation increases DNA uptake by the exposure of protoplasts to high voltage; small pores are made in the cell membrane.
- Conjugation, i.e., cell-to-cell contact, makes possible the unidirectional transfer of DNA from one bacterial cell to another.
- Bacteriophages (viruses that infect bacteria) are used to carry DNA into bacterial cells; this process is called transduction.

EXPLANATORY

DNA Hybridization or Homology

This procedure is used to determine the degree of similarity of DNA from different sources. The procedure involves the following steps:

- The double-stranded DNA molecules from two organisms, e.g., bacteria are heated to convert them to single strands.
- The single DNA strands from each organism are mixed and allowed to cool.
- The strand of one organism is labelled to enable detection of hybridization.
- The extent to which double strands are formed (heteroduplexes) indicates the degree of relatedness of the two organisms.

This procedure is particularly useful in the taxonomy (classification) of bacteria and other microorganisms at the species level.

Genomic Library (Gene Bank)

This is a collection of recombinant clones that contain all of the DNA present in an individual organism. Preparation of a genomic library involves the following steps:

- Cut up the entire genome with one or more restriction enzymes
- Splice as many as possible of the DNA fragments into copies of a vector, e.g., plasmid, phage.
- Introduce recombinant vectors into bacteria, often *E. coli.*
- Various procedures are used to identify the clone with the target sequence: these include hybridization with a labelled DNA probe, followed by autoradiographic screening for the probe label; or immunological screening for the protein product; and screening for protein activity.

There is no order to the clones of a library. When needed the relationship among the clones is established by the rather involved procedures of physical mapping.

Plasmids

These are small circular DNA molecules found mainly inside bacteria (see Figure 4.2) but not integrated into the genome. Some are species specific while others can be transferred to other bacterial species (broad host-range plasmids). The plasmids replicate everytime the cell reproduces its own chromosome. They generally are not essential for cell survival, and most, but not all bacteria have them.

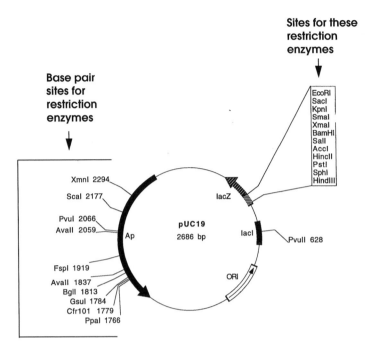

Figure 4.2. Bacterial plasmid pUC19; size 2686 base pairs. Genes are: *Ap* (ampicillin resistance), *lacZ* (ß-galactosidase), and *lacI* (ß-galactosidase repressor.) Ori indicates where plasmid replication begins. Reference: Yanisch-Perron, C., Vieira, J., & Messing, J.: Gene, 33:103-119. 1985.

Plasmids are uncommon in eukaryotic cells. The most studied eukaryotic plasmid is that which occurs in baker's yeast. The plasmid is called "2μm circle" and is used as a vector for cloning genes.

Plasmids carry various determinants, i.e., genes, which can contribute to the survival of the host bacteria., e.g., genes for antibiotic resistance. They can also be used as vectors to shuttle recombinant genes into other species.

Plasmids have from one to several hundred genes. Some have antibiotic resistance genes. Antibiotic resistance may serve as a selectable marker in gene cloning to ensure that bacteria contain a particular plasmid.

Plasmids range in size from 1 kb to 250 kb; those of 10 kb or less are often used as cloning vehicles.

Copy number refers to the number of copies of an individual plasmid found in a bacterial cell; it usually ranges from 1 to 50; multiple copies are preferred for cloning. There may be several different kinds of plasmids in a cell. There are five main types of plasmids:

- Fertility or F plasmids carry transfer genes which mediate conjugal transfer.
- Resistance or R plasmids carry genes conferring on the host bacterium resistance to antimicrobial agents.
- Col plasmids code for colicins (proteins) that kill some bacteria.
- Degradative plasmids code for enzymes that allow the host to break down certain compounds.
- Virulence plasmids, e.g., the Ti plasmids of the bacterium *Agrobacterium tumifaciens,* are involved in producing crown gall disease of broad-leafed plants.

The 2µm plasmid is a naturally-occurring, double-stranded, circular DNA plasmid (6,318 bp) found in the nuclei of a yeast cell (Plasmid Amplification and Plasmid Fingerprinting, see Glossary).

Vectors other than plasmids used in gene cloning are: bacteriophages, cosmids, and yeast artificial chromosome (YAC).

Restriction Endonuclease

This is an enzyme that cuts DNA molecules only at a limited number of specific nucleotide sites (see Figure 4.3). The natural function of these enzymes is to protect the cell against invasion by foreign DNA.

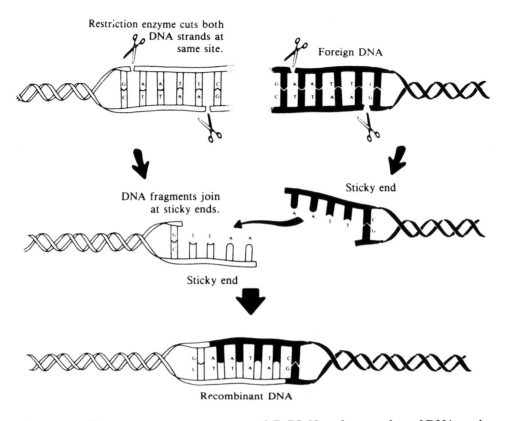

Figure 4.3. The restriction enzyme activity of *Eco*RI. Note that stretches of DNA cut by the restriction enzyme are palidromes, i.e., they read the same in either direction. Restriction enzymes require this kind of base pair symmetry. The cut is made between G and A in one strand and A and G on the other. The foreign DNA bonds with the original DNA producing recombinant DNA. From *Genethics* by D. Suzuki and P. Knudson. Copyright 1989 by New Data Enterprises and Peter Knudson. Reprinted by permission of Harvard University Press.

Several hundred restriction enzymes have been identified from   a variety of bacteria. Those used in cloning are type II; type I and type III are more complex enzymes and not important in gene cloning. These enzymes are named for their bacterial origin, e.g., *Eco*RI is from *Escherichia coli* strain RI; the E is from the genus *Escherichia*; co from the species name *coli*; R is for the strain; and I for the first restriction enzyme from *E. coli* .

The *Eco*RI enzyme is able to recognize and bind the sequence GAATTC. There is a recognition sequence for each of these enzymes, e.g.,   *Eco*RI: 5'GAATTC 3'
                                 3'CTTAAG 5'

Almost all recognition sequences are palindromes (sequences that read the same in both directions). (See Table 4.1, Recognition Sites for Some Commonly Used Restriction Endonucleases.)

There are more than 400 restriction enzymes capable of cutting DNA to produce more than 100 different sequences of bases.

Table 4.1. Restriction Sequences for Some Commonly Used Restriction Endonucleases.

Bacterium	Enzyme	Recognition Sequence	Blunt(B) or Sticky (S) end
E. coli	EcoRI	5^1...G$^\downarrow$A-A-T-T-C...3^1 3^1...C-T-T-A-A$_\uparrow$G...5^1	S
Staph. aureus	Sau3A	5^1...$^\downarrow$G-A-T-C...3^1 3^1...C-T-A-G$_\uparrow$...5^1	S
Proteus vulgaris	PvuI	5^1...C-G-A-T$^\downarrow$C-G...3^1 3^1...G-C$_\uparrow$T-A-G-C...5^1	S
P. vulgaris	PvuI I	5^1...C-A-G$^\downarrow$C-T-G...3^1 3^1...G-T-C$_\uparrow$G-A-C...5^1	B
Thermus aquaticus	TaqI	5^1...T$^\downarrow$C-G-A...3^1 3^1...A-G-C$_\uparrow$T...5^1	S

NOTE: Arrows indicate sites of cleavage.

Chapter 5

RESTRICTION FRAGMENT LENGTH POLY-MORPHISMS (RFLPs), AND VARIABLE NUMBER TANDEM REPEATS (VNTRs)

Restriction Fragment Length Polymorphisms (RFLP) and Variable Number Tandem Repeats (VNTR) reveal many differences in DNA sequences among individuals of a species. RFLP and VNTR involve cleavage of chromosomal DNA by restriction enzymes, separation of the fragments by gel electrophoresis (see Figure 5.1) and visualization of bands (DNA fragments) by staining with fluorescent ethidium bromide (see Figures 5.1 & 5.2). An important use of RLFPs and VNTRs is referred to as DNA or Genetic Fingerprinting. It is employed in the identification of animals and humans, and in taxonomic and epidemiologic studies. On the average one base pair in 1000 is different between unrelated individuals. DNA fingerprinting makes possible the identification of varieties of microorganisms with great precision and the reliable, positive identification of individual humans and animals. Restriction enzymes thus have many practical applications.

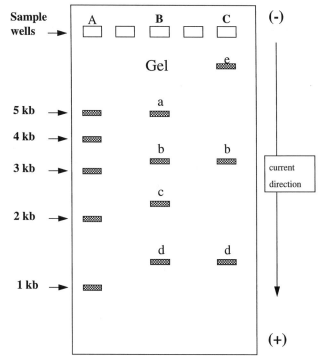

Lane A: marker DNA, fragments of known size

Lane B : DNA sample #1: cutting with restriction enzyme yields 4
fragments: (a,b,c,d)

Lane C: DNA sample #2: cutting with restriction enzyme yields 3 fragments
because of base change

Figure 5.1. Separation of DNA fragments by gel electrophoresis. An electric current is applied after the DNA has been added to the wells at A, B, and C. The DNA fragments move at different speeds depending on their size. Ethidium bromide added to the gel stains the bands a, b, c, d, and e, and allows them to be seen under ultraviolet light.

RESTRICTION FRAGMENT LENGTH POLYMORPHISMS (RFLPs)

As mentioned above, RFLPs are variations in the lengths of certain DNA fragments produced after cleavage with a restriction endonuclease (see Figure 5.2). Variations occur particularly in the "junk" stretches of DNA between the genes. Apparently meaningless codon triplets may be repeated over and over again. In some of these stretches there is a difference between individuals. The differences in the length of the DNA segments are due to the presence or absence of a specific restriction endonuclease recognition site(s). When a single base pair change occurs at a homologous site in an allele (an alternative form of a gene) the sequence is no longer a recognition site and the DNA is not cut; thus one allelic form has a functional site for a particular restriction endonuclease, and the other does not. The differences in the lengths of DNA after enzyme treatment are determined by gel electrophoresis; however, there may be so many fragments that it is not possible to distinguish a single altered band (see Figure 5.3).

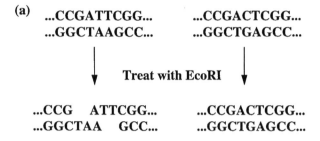

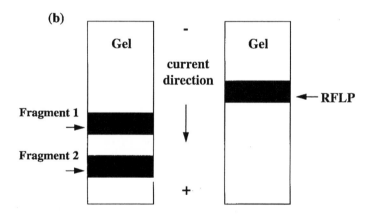

Figure 5.2. Restriction Fragment Length Polymorphisms (RFLP). (a) The two DNAs are from different people. When these sample DNA pieces are treated separately with the enzyme *Eco*RI, the one on the left will break into two fragments because it has the sequence that EcoRI cuts; the single base change in the DNA on the right has changed the sequence so that the site for cutting with the enzyme is not present. The piece of DNA on the left will yield two fragments, the one on the right only one, the RFLP. (b) When the fragments are placed in separate lanes of a gel through which an electric current is passed size differences are revealed. The two short fragments in the left-lane move farther down the gel than the long fragment. Thus the two individuals differ in this particular region of their DNA. Different alleles have been identified at a RFLP locus. Adapted from *The Lives to Come* by P. Kitcher. Simon & Schuster, New York, N.Y., 1996.

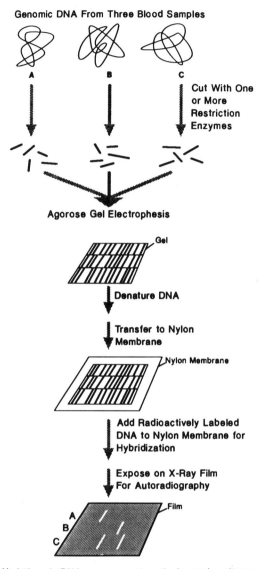

Variations in DNA sequences at particular marker sites are observed as differences in numbers and sizes of DNA fragments among samples taken from different individuals (shown here as samples A, B, and C).

Figure 5.3. The detection of Restriction Fragment Length Polymorphisms (RFLPs) using radioactively labelled DNA probes. Reprinted from: U.S. Congress, Office of Techology Assessment, *Mapping Our Genes—The Genome Project: How Big, How Fast?* OTA-BA-373 (Washington, D.C.: U.S. Government Printing Office, April 1988.)

In such cases, the DNA bands in the gel are transferred to a nitro-cellulose or other membrane (Southern blotting or Transfer). After Southern Blotting the different restriction endonuclease fragments are detected by hybridization to a labeled DNA probe (one that spans all or part of the region of interest).

The probe will hybridize only to the restriction fragments in the relevant region, and the sizes of these fragments will be seen following development of the labeled probe (e.g., autoradiograph) (see Figure 5.3). These sizes are compared with the pattern for the unmutated gene and a RFLP, if present, will be clearly visible. As mentioned above, this technique of DNA or Genetic Fingerprinting may be used to determine if DNA samples belong to the same or different individuals. RFLP analysis requires the DNA from about 1000 cells. The technique is very accurate but relatively expensive.

Known mutations are rare, but RFLPs are plentiful and for this reason, they are helpful in constructing the human genetic map (see RFLP Linkage Analysis, Chapter 9).

In forensic DNA testing, the restriction endonuclease HaeIII is frequently used; the recognition sequence is GGCC with blunt ends. As many as five different DNA probes may be used.

RLFP analysis has many other applications where the identity of individuals is involved, e.g., in immigrant disputes and in paternity and maternity determination.

A probability figure is calculated to express the likelihood of more than one person having exactly the same DNA banding pattern.

What is referred to as the LOD score is the logarithm of the ratio of the odds for or against linkage of two markers. It is used in RLFP linkage analysis. The higher the LOD number, the closer together are the two genetic markers, i.e., two RLFPs, on the chromosome likely to be. If the LOD score is three the odds are 1000 to one that the linkage was not due to chance.

VARIABLE NUMBER TANDEM REPEATS (VNTRs)

VNTRs can be considered a particular type of RFLP. There are different DNA sequences at restriction sites of "junk" DNA; these characteristic sequences repeat themselves a different number of times in different individuals; this phenomenon is called Variable Number Tandem Repeats (see Figure 5.4).

Repeated Regions Within DNA
of Two Different Persons

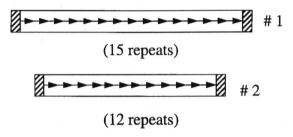

Figure 5.4. Variable Number Tandem Repeats (VNTRs). Regions of human DNA vary in the number of short, repeated regions. The DNA from person 1 has 15 repeats whereas that from the second person has 12. Outside these repeats are sequences (hatched) recognized as cutting sites by particular restriction endonucleases. Cutting within those sites (arrows) produces fragments of different lengths for the two people. Adapted from *Doube-Edged Sword* by K. Drlica. Addison-Wesley Publishing Company, New York, N.Y., 1994.

Differences in VNTRs are inherited with the same stability as other alleles so every length of a simple repeat can be treated as a different allele. Only identical twins have identical DNA fingerprints, and only individuals from very inbred groups are likely to have more than a few VNTRs in common.

VNTRs are used for DNA fingerprinting and the analysis is quicker and less expensive than that using RFLPs. Determination of VNTRs involves the following:

• DNA is first isolated from the samples.
• DNA is then cut with a restriction endonuclease that cuts outside the VNTR; this results in fragments of different length, e.g., one person might have a VNTR with 12 repeated segments and another person with a VNTR with 20 repeats.
• The fragments are then subjected to gel electrophoresis but as with RFLP analysis, there are so many different fragments of varying size that a transfer is made to a nitrocellulose or nylon membrane (Southern Blotting).

• Labeled probes that bind only to the DNA in bands of interest are applied to the membrane; development of the labeled probes will indicate to which bands the probes bind.

The odds for finding a positive binding pattern between two DNA samples using VNTR analysis increases with the number of probes used.

Amplifying three or four VNTRs by the Polymerase Chain Reaction (see Chapter 6) and separating them by size in a gel is a rapid and reliable way to get a record of bands unlikely to appear by chance.

Figure 5.5 illustrates a forensic application of VNTR analysis.

Variable Number Tandem
Repeats (VNTRs)

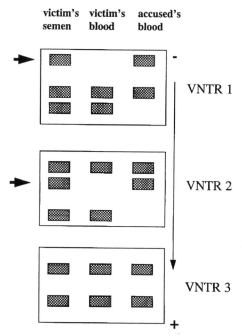

Figure 5.5. DNA banding pattern used for DNA fingerprinting. Restriction fragments are separated by gel electrophoresis then transferred to a membrane for easier recognition (Southern Blotting). Labeled DNA probes bind to specific VNTR restriction fragments. Three different VNTRs were probed. The arrows indicate bands attributed to the accused rapist and strongly indicate his guilt. Adapted from *Doube-Edged Sword* by K. Drlica. Addison-Wesley Publishing Company, New York, N.Y., 1994.

Chapter 6

POLYMERASE CHAIN REACTION (PCR)

POLYMERASE CHAIN REACTION (PCR)

PCR is a procedure for amplifying small sections of DNA (200 to 2000bp) by binding DNA primers (see below) to sections of the DNA to be amplified. Except for the RAPD procedure described below, it is necessary to know, at least approximately, the short sequences (20-30bp) of the DNA segment or gene being sought in order to generate the primers. The region to be amplified can be up to 10,000 bp, but generally most are 200 to 2000 bp. Oligonucleotide synthesizers are semiautomated machines available for the synthesis of primers.

Repeating cycles of heating and cooling, in the presence of thermostable DNA polymerase (*Taq* polymerase–see below) result in the creation of new sections of DNA between the primers hybridizing to specific sequences. The various steps in the procedure (see Figure 6.1) are:

- The sample of unknown DNA is heated (90°C) so the double-stranded DNA splits apart.
- While cooling, a mixture of primers (short nucleotide strands complementary to a region of the DNA being sought) is added. The primers bind to the regions of complementarity on the DNA single strands.
- The temperature is lowered (55°C) and a special DNA polymerase (*Taq* polymerase) is added to the mixture along with an abundance of unattached nucleotides; the polymerase brings the nucleotides to the DNA chain where they are attached in order and reflecting the complementary nucleotides.

- When all of the DNA chain is constructed between the primers, a complete copy of the targeted DNA sequence is obtained, i.e., double stranded DNA (dsDNA).
- The temperature is raised again to split the dsDNA and each one makes two copies of single-stranded DNA; after cooling, each copy serves as a target strand for the primers and *Taq* polymerase for the next round of DNA synthesis; after many cycles (e.g., 1-3 hours), millions of copies of the target sequence are obtained.

The heat resistance (90°C) of the special DNA polymerase makes PCR possible.

(a) Specific region of DNA strand separated by heating (denaturing)

(b) Primers (2 short oligonucleotides) annealed to denatured DNA to delimit region to be amplified

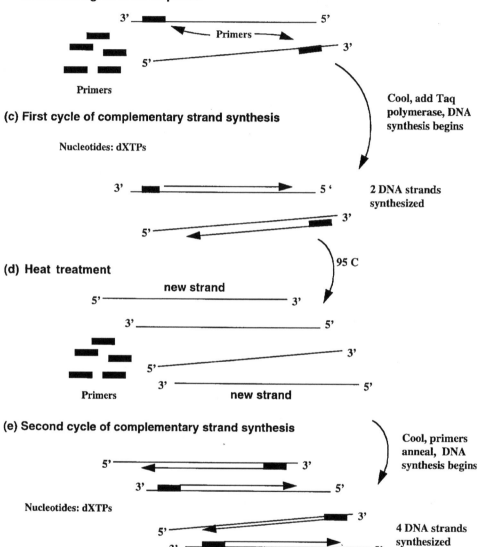

(c) First cycle of complementary strand synthesis

(d) Heat treatment

(e) Second cycle of complementary strand synthesis

(f) Third cycle results in 8 strands; fourth in 16 strands and so on.

Figure 6.1 Polymerase Chain Reaction (PCR).

PCR is of great value in many crime cases in which the amount of DNA available is too small for RFLP/VNTR analysis. For various reasons, including the age and degradation of the sample, the amount of DNA at a crime scene may be very small. The tiny amount is amplified by PCR so that there is sufficient DNA to perform RFLP/VNTR analysis (see Figure 5.3). Only a few dozen cells are required for PCR and it can be used even if the DNA is somewhat degraded. However, it is more precise and expensive than RFLP analysis.

A new PCR protocol has been developed for forensic use that compares short triplet repeat sequences in as many as 13 locations in human DNA. These are sites where the number of repeats varies widely among individuals. A number of primers are used which yield fragments of different size that can be separated by gel electrophoresis. It is quicker to perform than RFLP. The odds are 1 in 100,000 that individual DNA samples match by coincidence.

PCR has a number of applications in addition to those mentioned, including genetic testing (detection of mutations), identification of individuals, and the laboratory diagnosis of diseases caused by bacteria, viruses, fungi, and protozoa (see Chapter 10). All that is needed is knowledge of enough of the DNA sequence involved and suitable primers. PCR can be readily applied to the diagnosis of a number of viral diseases in that their nucleotide sequences are known.

The great advantage of PCR in microbial diagnosis is its usefulness when the amount of DNA is very minute (presence of virus in 1 in 1,000 human cells) and present in dead cells. PCR also provides sufficient specific DNA to make possible identification of the microorganism not possible by conventional laboratory means.

PCR has been adapted for the detection of amplified DNA products by enzyme immunoassay (ELISA). This is useful for screening large numbers of clinical specimens when looking for microbial pathogens. In doing PCR great care must be taken to exclude even minute amounts of contaminating DNA (e.g., from microorganisms associated with dust particles).

RANDOM AMPLIFICATION OF POLYMORPHIC DNA (RAPD)

This is a recent DNA polymorphism assay based on the PCR amplification of random DNA fragments using single short primers made up of arbitrary nucleotide sequences (arbitrary primers).

Prior sequence information is not required. The polymorphisms have been termed random amplified polymorphic DNA markers. The amplified products are resolved by gel electrophoresis.

RAPD can be used to evaluate the genetic relatedness among microorganisms of different genera and species. For example, this fingerprinting procedure is capable of detecting small oligonucleotide differences representing antigenic variation among strains of the same bacterial species.

Multiplex Polymerase Chain Reaction (MPCR) (This procedure is described in Chapter 9.

EXPLANATORY

Primers

These are short stretches of DNA or RNA used as starting points for nucleic acid synthesis. A primer hybridizes with a template strand of nucleic acid and provides a 3' hydroxyl end for the initiation of synthesis. The primers delimit the region that will be amplified. In PCR, two (sometimes more) synthetic oligonucleotide primers (about 20 nucleotides each) complementary to regions on opposite strands flank the target sequence; the 3' hydroxyl ends are oriented to each other (see "prime" in the Glossary). The target sequence in a sample is about 100 to 2000 bp in length. Designed primers and arbitrary ("off-the-shelf") primers are used.

Taq Polymerase

This is the DNA polymerase used in PCR. It is derived from the bacterium *Thermus aquaticus* that lives in hot springs; the polymerase's thermostability makes PCR possible.

Chapter 7

GENE-MAPPING

Anumber of strategies have been, and are being used for gene-mapping. They are outlined briefly below. (Figure 7.1 illustrates overall genetic mapping.)

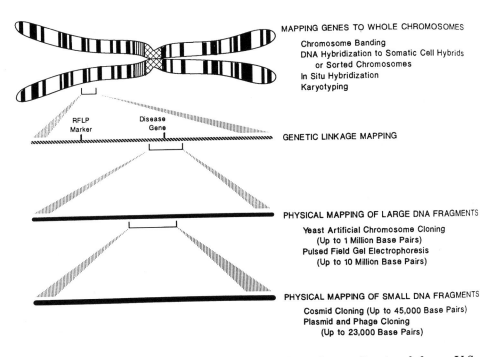

Figure 7.1. Genetic mapping at different levels of resolution. Reprinted from: U.S. Congress, Office of Technology Assessment, *Mapping Our Genes—The Genome Project: How Big? How Fast?* OTA-BA-373 (Washington, D.C.:U.S. Government Printing Office, April 1988).

51

CONTROLLED BREEDING EXPERIMENTS

These experiments led to the discovery of gene linkage, i.e., the tendency for nonallelic, neighboring genes located on the same chromosome to be inherited together.

As the distance between the two genes increases, the greater is the likelihood that the two genes will separate due to meiotic crossing over. The experimental crossover frequencies of hereditary characteristics made possible the drawing of crude linkage maps showing the linear order of some individual genes on chromosomes. Unfortunately, such maps only revealed relative spatial relationships between genes.

FAMILY PEDIGREES

The study of pedigrees or family histories of large stable families (e.g., the Amish) has been helpful. This has involved the tracing of a particular trait through several generations of a family tree.

Patterns of inheritance can sometimes be correlated with patterns of chromosomal inheritance and make possible determination of the chromosomal location of a gene. Chromosomal banding patterns have been helpful in this process.

MAPPING GENES TO WHOLE CHROMOSOMES

Chromosome Banding

When chromosomes are appropriately stained they display bands which correspond to regional differences in the density of adenine-thymine and cytosine-guanine pairings (see Figure 7.2).

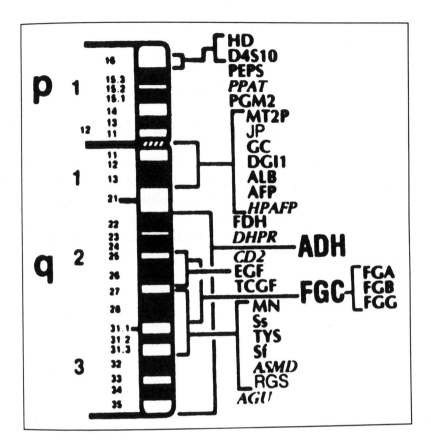

Figure 7.2. Pictorial representation of a human gene map. The banding patterns on the chromosome are produced in the laboratory with special staining techniques. Each chromosome has two arms, a short or "p" arm and a long "q" arm. The different divisions in each arm are numbered, with "1" starting at the central division point or centromere and the largest number just before the outer end or "telomere" of each arm. The rough location of different genes or genetic markers is indicated by the abbreviations. For example, on chromosome 4 the abbreviations "HD" and "D4510" at the outer end of the p arm stand for genetic markers associated with Huntington's disease. From *Mapping the Code* by J. Davis, John Wiley & Sons, Inc. New York, N.Y. 1990.

There is some variation in the banding pattern of the 23 homologous chromosomes from one person to another. This variation is called chromosomal fingerprinting and can be used to identify individual chromosomes.

There are about 10 bands per chromosome and each band has about 10 to 12 million base pairs. The pattern of bands serves as stable chromosomal landmarks. Centromeres also serve as landmarks. Variation in chromosomal banding can be due to loss (deletion) or duplication of a chromosomal fragment; either can be associated with a genetic disorder.

Somatic Cell Hybridization

This technique is useful in obtaining single chromosomes for analysis, including sequencing.

Human tumor cells are fused with those of another species, e.g., mice (see Figure 7.3). Certain chemicals, an electric field or Sendai virus enhance cell to cell fusion in cell cultures. The hybrids of, e.g., human and mouse cells are unstable and end up with about eight to 12 of the 46 human chromosomes in addition to the remaining mouse chromosomes.

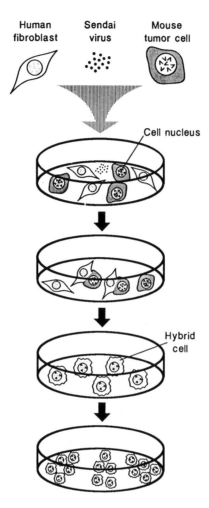

Figure 7.3. Somatic Cell Hybridization. Reprinted from: U.S. Congress, Office of Technology Assessment, *Mapping Our Genes*–The Genome Project: How Big, How Fast? OTA-BA-373 (Washington, D.C.: U.S. Government Printing Office, April 1988).

On subculture, hybrids may be obtained with only one human chromosome. Hybrid identity can be determined by its chromosomal pattern.

The flow cytometer (see Glossary) can be used to obtain individual chromosomes from mixtures of chromosomes.

In Situ Hybridization

This involves denaturing DNA, e.g., of a chromosome inside of the cell nucleus, then applying a probe consisting of a cloned gene. When hybridization occurs between the cloned gene and its chromosomal copy, a specifically labeled segment can be seen on the chromosome with the aid of a microscope.

Karyotyping

Karyotyping involves determining the number, size, and shape of the set of chromosomes of a cell. This is done by examining photomicrographs of chromosome preparations. Karyotyping has been used to determine if there is a correlation between chromosomal abnormalities and features of particular diseases, e.g., Down syndrome.

RESTRICTION MAPPING

Restriction mapping is sometimes referred to as genetic linkage mapping. Restriction maps using RFLPs incorporate information gathered from the genetic inheritance patterns of large multigenerational families. The search is for a DNA segment, an RLFP, that is inherited with the genetic disorder. Restriction maps have a resolution about ten times higher than that of cytogenic maps (those based on banding variation). One can identify a gene location to within a million base pairs.

PHYSICAL MAPPING OF LARGE DNA FRAGMENTS

Finding the location of a cloned gene on a large DNA molecule:

Restriction mapping becomes more difficult with molecules larger than about 240 kb. Molecules larger than about 60 kb cannot be separated efficiently by conventional gel electrophoresis; however, they can be separated by pulsed field gel electrophoresis (PFGE). Molecules as large as several thousands of kb can be resolved with PFGE, including chromosomal molecules of yeast, fungi, and many other eukaryotes.

The DNA of individual chromosomes can be purified by PFGE making possible chromosomal gene libraries. Each of these libraries is smaller and thus easier to manage than the complete genomic library.

Southern blotting can be used on the chromosomal DNA separated by PFGE, followed by hybridization with the appropriate probe and thus making possible the identification of a specific gene.

PHYSICAL MAPPING OF SMALL DNA FRAGMENTS

Finding the location of a gene on a small DNA molecule:

The location of a cloned gene on a small DNA molecule such as a plasmid can be determined as follows:
- Gene x has been cloned in plasmid y.
- The DNA of plasmid y is treated by a restriction endonuclease, then subjected to gel electrophoresis.
- The fragments are separated yielding, e.g., 13 bands; one of these fragments will be the same as that inserted in plasmid y.
- The DNA bands are transferred to a nitrocellulose membrane (Southern Blotting).
- A labeled probe (the recombinant molecule carrying gene x) is applied, hybridization will occur and probe visualization will show which restriction fragment contains gene x.
- This makes it possible to determine the position of gene x on the plasmid y restriction map.

Cosmids (accommodating up to 45,000 base pairs) and bacteriophages (up to 23,000 bp) are also used for mapping small DNA fragments.

CHROMOSOME WALKING AND CHROMOSOME JUMPING

These techniques are also used to find the position of a cloned gene on large DNA molecules. They are rather complex procedures which are defined briefly as follows:

Chromosome Walking:

A strategy in which a nucleotide sequence near one end of a cloned region is used as a probe for locating adjacent, overlapping regions.

Gene walking enables very long and very dense maps to be drawn-up; it was used to map the genome of *E. coli* (4.4 million base pairs).

Chromosome Jumping:

In this procedure two segments of double-stranded DNA that are separated by about 200 kb are cloned together (circularized).

After subcloning, each segment is used as a probe to identify the cloned DNA sequences that are about 200 kb apart.

CONTIG MAPPING

This rather complex procedure utilizes groups of clones in a library, which represent overlapping or contiguous regions, to determine the location of common subfragments.

Chapter 8

DNA SEQUENCING AND THE HUMAN GENOME PROJECT

DNA SEQUENCING

The ultimate mapping is DNA sequencing. It involves the determination of the precise order of nucleotides in a fragment of DNA. The two principal sequencing techniques are the Chain Termination Method (Sanger and Coulson) and the Chemical Degradation Method (Maxam-Gilbert). The former, which is used most frequently, is outlined below (see Figure 8.1).

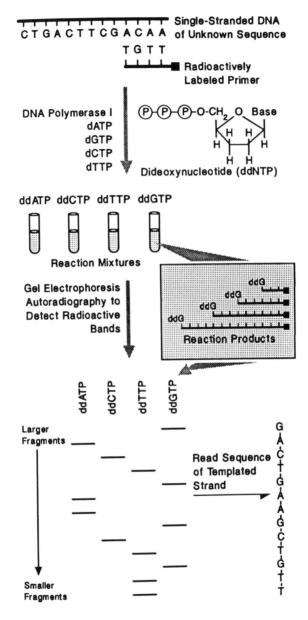

Figure 8.1. The Sanger method of DNA sequencing. Reprinted from: U.S. Congress, Office of Technology Assessment, *Mapping Our Genes—The Genome Project: How Big, How Fast?* OTA-BA-373 (Washington, D.C.: U.S. Government Printing Office, April 1988).

DNA is analyzed by allowing copies (obtained usually by cloning in M13 bacteriophage or in a YAC) of an undeciphered strand of DNA to replicate artificially in four separate test tubes:

- Each test-tube solution contains four deoxynucleotides designated A, C, G, and T plus a low concentration of a chemical analogue (dideooxynucleotide) of one of these nucleotides, e.g., G, ddG and the enzyme DNA polymerase that initiates DNA synthesis from the end of the sequence, and a primer which the DNA polymerase extends by incorporating DNTPs.
- The analogue is a chain terminator; polymerase will add it to the growing chain, but the next nucleotide will not attach to it.
- Whenever by chance a G analogue is incorporated instead of G, the synthesis stops.
- Thus when the reaction is completed, the solution contains only DNA chains that end in the G analogue.
- There are four solutions in all containing DNA chains, each tube containing only chains that end in one nucleotide.
- There will be chains of every length from one nucleotide up to the whole original fragment if no analogue was incorporated.
- The four solutions are subjected to gel electrophoresis (a thin poly acrylamide gel is used) in four lanes; the smallest chains will migrate the most rapidly.
- (A radioactive nucleotide was added earlier to the reaction mixtures so that autoradiography could be used to visualize the bands.)
- When visualized the one-link chain is therefore at the one end of the gel, and if it is in the lane with the A analogue solution, the first nucleotide in the sequence is A.
- If the chain that moved the next farthest (one nucleotide longer) is in the T analogue lane, the next nucleotide in the sequence is T; the sequence this far is AT.
- As one moves along the lanes, the sequence is added to, e.g., the third band was ATddT (dd stands for the analogue).
- The reading is continued until the bands become too dense to distinguish one from another.

About 400 nucleotides can usually be read from one autoradiograph. An automated procedure for DNA sequencing is described below.

In the Maxam-Gilbert method, single-stranded DNA is obtained from double-stranded DNA. The former is labelled at the 5' end with phosphorus 32. By chemical cleavage procedures, breaks are made on one side of a particular base and fragments are separated according to size by gel electrophoresis. They are then identified by autoradiography.

HUMAN GENOME PROJECT

There is an international organization, The Human Genome Organization, which is assigned to coordinate the Human Genome Project. The objective of this project is to sequence all of the bases in the human genome.

The first phase of the Human Genome Project involves mapping as described earler. The positional markers determined by mapping, e.g., by "restriction mapping," are useful in finding where in each chromosome genes are located.

The second phase involves DNA sequencing; each laboratory involved in the project works on a particular chromosome or section of a chromosome. Millions of copies of DNA fragments are required for sequencing; these are provided by recombinant DNA procedures described above (Cosmid and YAC libraries) and below.

Several methods have been developed for automating some of the procedures for DNA sequencing. They are based on either the Maxam and Gilbert method or that devised by Sanger (see above).

AN AUTOMATED PROCEDURE FOR SEQUENCING
NUCLEOTIDES OF DNA

Automated procedures have greatly accelerated sequencing in the human genome project. One such procedure, based on the Sanger method, is summarized below:
   • select a fragment of human chromosomal DNA to be sequenced.
   • both introns and exons are sequenced.
   • DNA fragments are obtained by treatment with restriction endonuclease.
   • the fragments are inserted into a special plasmid (see Insertional Inactivation, Chapter 8) which in turn is incorporated into bacteria (transformation).

- the bacteria are plated on a special medium (containing ampicillin, X-gal–a lactose analogue–and IPTG, the inducer of an enzyme ß-galactosidase).
- bacterial colonies appear after incubation.
- those colonies containing a DNA fragment appear white; those that are blue are devoid of the DNA fragments.
- robots scan the petri dishes for white colonies picking and transferring them to small wells in a plate; the plates are incubated to obtain further growth.
- a few hours later, the plasmids are extracted and specific cloned fragments are amplified several million times by PCR (see Chapter 6).
- the fragments are then denatured, thus providing single-stranded segments a few hundred nucleotides in length.
- fluorescent dyes which yield different colors are used to tag the dideoxynucleotide bases thus: cytosine = blue, guanine = yellow, adenine = green, and thymine = red.
- four different tagged dideonucleotide bases are incorporated into the ends of the different DNA segments.
- the tagged fragments are separated by gel electrophoresis.
- at the bottom of the tube containing the gel a laser beam excites the dye in the tagged fragment and causes it to fluoresce; a photo-electric cell detects the color responses which are recorded by a computer.
- the computer collates the data and provides the nucleotide sequence.

An automated DNA sequencing procedure using fluorescently labeled DNA is sketched in Figure 8.2.

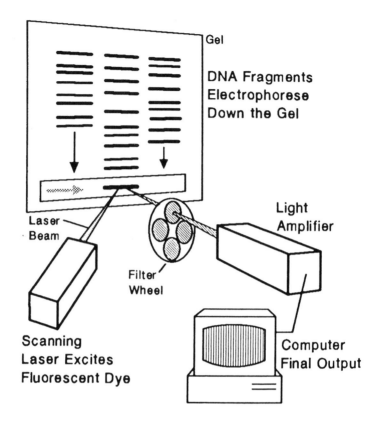

Figure 8.2 Automated DNA sequencing using fluorescently labeled DNA. Source: Leroy Hood, California Institute of Technology, Pasadena, CA. From U.S. Congress, Office of Technology Assessment, *Mapping Our Mapping–The Genome Project: How Big, How Fast?* OTA-BA-373 (Washington, D.C.: U.S. Government Printing Office, April, 1988).

Although the human genome has not yet been completely sequenced, the genomes of a yeast, a fruitfly, a nematode, and several bacterial species have been sequenced. Single-celled organisms called Archea, some of which have been isolated from hot springs, are now no longer considered to be bacteria. The genome of one Archea species has been sequenced and found to be more closely related to eukaryotes than bacteria. In fact, it is conjectured that the Archea were the ancestors of eukaryotes.

GENBANK

GenBank is one of several data bases and repositories for genetic information in the United States which provides nucleotide sequence data. Other centers make available genetic map data, protein and amino acid sequence and structure data, and human DNA probe and chromosome libraries. There was a vast increase in sequenced DNA when it was found in the 1970s that genetic maps could be created using special markers within chromosomes.

GenBank is a federally-funded database located at the National Center for Biotechnology in Bethesda, Maryland. It is a repository for DNA sequences involving millions (652 million as of October, 1996) of bases from many organisms including viruses, bacteria, plants, fungi, fruit flies, protzoa, nematodes (small round worms), mice, chimpanzees, and humans. Of the millions of bases recorded, about a quarter are human.

The GenBank is useful in a number of ways. The relatedness of genes in the various genomes can be determined by homology searches using computers and search algorithms (systematic procedure for carrying out a computation). It should ultimately be possible to trace all of the 100,000 or so human genes back to a small number of progenitor or ancestral genes. This will contribute greatly to understanding evolution and the genetic relationships among various organisms.

The use of algorithms to determine the relatedness of amino acid sequences has also made it possible to compare "new" proteins to those that have already been identified either through DNA sequencing or X-ray crystallography, and thus aid in determining their three-dimensional structure and function.

CHEMICAL SYNTHESIS OF DNA

DNA synthesizers or "gene machines" are now available to automate the chemical reactions involved in DNA synthesis. The method of choice at present in the Phosporamide Method. It involves many steps and those interested are referred to texts on Molecular Biotechnology for details.

Chemically synthesized, single-stranded DNA oligonucleotides are now used to assemble gene fragments and whole genes. Probes can be produced to screen gene libraries, and several kinds of single- and double-stranded oligonucleotides can be synthesized to facilitate gene cloning.

LIMITATIONS OF THE HUMAN GENOME PROJECT

Lewontin (*Biology as Ideology*, Harper Collins, New York, 1993) has raised the following points regarding the Human Genome Project:
- There are about three billion nucleotides in the human genome and any two individuals will differ in about 600,000 nucleotides.
- An average gene has about 3,000 nucleotides; the difference between the genes of two normal persons would be about 20 nucleotides.
- Whose genome will represent the normal person?
- The average person has a large number of defective genes covered by a normal copy from the other parent, thus there will be many unknown defective genes entered in the catalogue.
- When the DNA from a person with a genetic disorder is compared with the standard normal sequence provided by the the human genome project, it would be impossible to decide which, if any, of the many differences between the two DNAs is responsible for the disease.
- It would be necessary to look at a large population of diseased and normal people to determine if there are common differences; even this may not solve the problem if the disorder has a multiple genetic cause.

EXPLANATORY

Insertional Inactivation

A cloning procedure in which insertion of a new piece of DNA into a gene carried by the vector (often a plasmid) inactivates that gene. In the automated procedure described above, the recombinant plasmid molecule has an ampicillin resistance gene and the new DNA is inserted into the *Bam*HI (restriction endonuclease) site; this inactivates the *lacZ'*gene. The latter gene is responsible for encoding an enzyme which breaks down the lactose analogue X-gal.

Bacteria with both recombinant and nonrecombinant plasmids are ampicillin resistant and thus will grow on the special medium. Because of the inactivation of the *lacZ'* gene, the bacteria containing the recombinant molecule cannot produce ß-galactosidase that breaks down X-gal in the culture medium and thus bacterial colonies appear white.

Those bacteria containing plasmids in which the new DNA has not been inserted in the *Bam* HI site have an active *lac* Z' gene resulting in the breakdown of X-gal; the colonies appear blue on the culture medium.

Chapter 9

DEFECTIVE GENES AND GENETIC DISORDERS

Great efforts are being made to find the defective genes that predispose or lead to many important genetic diseases. Location and identification of these genes and knowledge of the mechanisms by which they cause disease are the basis for instituting ways of prevention and developing gene therapies (see Chapter 15) and treatment.

More than 200 inherited metabolic diseases can be detected by genetic tests, and a total of about 450 disorders have been linked to specific alleles. There are more than 4000 human genetic disorders for which the vast majority have not been associated with or defined by a specific allele.

Many genetic disorders also occur in both domestic and wild animals. Tests are available for the identification of genes of some important genetic disorders of domestic species.

MUTATIONS

Most genetic disorders are due directly or indirectly to mutations. It is thought that about 80 percent of single gene disorders are the result of mutations that occurred in the germ cells of distant ancestors. Sporadic mutations in the reproductive cells of one of the parents are thought to make up most of the remaining 20 percent.

Mutations are sudden, and often permanent changes that alter the sequence of nucleotide base-pairs on a DNA strand. The changes may range from alteration of only one base (point mutation) to the elimination of several base pairs or a complete gene (a deletion). If the sequence of the bases is sufficiently altered such that one or more amino acids of the protein encoded by the gene are changed, the protein may no longer be able to carry out its function. Mutations result in changes in the genotype (genetic composition). The phenotype (observable appearance) may or may not be changed.

Mutations occur naturally and randomly at a low rate, about one in a million. The mutation rate can be increased by beta and gamma irradiation, ultraviolet light, x-rays, and many chemicals (mutagens). Although most mutations are silent and some deleterious, a small number may be beneficial and have a role in evolution in helping select for fitness.

Induced mutagenesis is used routinely in research to obtain various phenotypes to allow for the study of gene structure and function, e.g., proteins with altered enzyme activity.

The inheritance of mutations leading to genetic disorders is discussed later under Genetic Screening/Testing.

Principal Mutations

Point mutation: a base substitution involving a single nucleotide; it may result in:
- The substitution of a different amino acid = missense mutation.
- No change in amino acid = silent mutation.
- Creation of a stop condon = nonsense mutation.

Frameshift mutation: a base pair is inserted or deleted; this usually affects mRNA. All triplets can be changed for the remainder of the sequence. The changes may include nonsense, silent, and missense mutations.

CHROMOSOMAL ABNORMALITIES

In addition to mutations involving genes, there are chromosomal abnormalities that lead to genetic disorders. These chromosomal abnormalities may be numerical or structural. The technique called Multiple Spectral Imaging (described below) is particularly useful for detecting chromosomal abnormalities. Examples of chromosomal abnormalities are provided below under "Some Human Genetic Disorders."

MITOCHONDRIA

A rare disease called Leber's optic neuropathy has been traced to a defect (single mutation) in mitochondrial DNA. Parkinson's disease and a form of diabetes have also been attributed to faulty mitochondria (see Mitochondria, Chapter 1).

RECESSIVE AND DOMINANT GENES

The effect on an individual of a recessive defective gene is masked by its correct partner, the dominant gene. The carrier of the recessive defective gene can pass it on to his or her progeny. In the case of the defective gene being dominant, it masks the correct partner gene and the individual inheriting the dominant defective gene develops the genetic disorder.

SEX-LINKED GENETIC DISORDERS

Some genetic disorders are sex-linked. In such diseases the mutant gene is located on the X chromosome; in females the gene is recessive when the partner gene is normal. As the male has only one X chromosome he will be affected if the X chromosome carries the mutant gene. Some examples of sex-linked disorders are given below.

SINGLE-GENE AND POLYGENIC DISORDERS

A single-gene disorder is caused by one gene, e.g., sickle cell disease.

A polygenic or multifactorial disorder results from the combined action of alleles of more than one gene, e.g., some cancers and diabetes. The recently detected "breast cancer gene" BRCA1 is considered to predispose for breast cancer. Cancer is discussed further below.

DNA REPAIR

Maintaining the integrity of the DNA within chromosomes of cells is imperative for the health of the organism. The chromosomal DNA of cells is constantly assaulted by external influences such as mutagenic chemical compounds and radiation. There is also assault from within the cell by by-products of normal cellular metabolism, e.g., oxidants produced during carbohydrate metabolism. The results of these assaults can be a change that converts one base pair into another or causes two adjoining bases to be cross-linked together. Such errors can lead to harmful mutations or inhibition of DNA replication; both consequences lead to loss of cell viability.

To prevent these errors, DNA polymerase has proofreading functions whereby newly synthesized DNA is scanned to edit-out base pair changes or cross-linkage and correct them. DNA repair is a complex subject and it is dealt with only briefly below. Various repair mechanisms are employed to restore the correct nucleotide sequence of a DNA molecule. These include:

• Enzymatic excision of defective single-stranded segments and replacement with newly synthesized segments; this is referred to as "cut-and-patch" repair.

• Excision and subsequent replacement of thymine dimers (formed by the linking of thymine bases) that result from ultraviolet and other radiation; a photoreactivating enzyme unlinks the dimers.

• "Nucleotide excision repair" is used to correct damage to DNA caused by various influences including those caused by internally- and externally-derived chemicals.

Bacteria have the capacity to repair their DNA but viruses do not. The greater mutation rate of viruses helps them elude the immune system of the host and thus contributes to their survival.

SEARCHING FOR DEFECTIVE GENES

Three of the approaches used to acquire information about the gene of interest are outlined below:

Probing for the Defective Gene

Development of a nucleic acid probe for genetic diseases requires a knowledge of the particular defect resulting in the disorder. An oligonucleotide probe can be prepared if the exact nature of the genetic defect is known. Hybridization to the probe takes place whether the defect is due to a deletion, mutation, or a point mutation.

A defect due to an insertion can be detected by a probe spanning the junction between the insertion and the gene.

The procedure for the direct identification of the mutated (defective) gene involves the following steps:

- The known mutated gene is cloned to provide sufficient copies for diagnostic purposes.
- A DNA probe is prepared (see Chapter 4) using a part (usually a sequence of 100 to 500 base pairs) of single-stranded DNA matching a portion of the known gene.
- A sample of DNA obtained from the individual to be tested is treated with a restriction endonuclease to produce fragments which are separated by gel electrophoresis.
- The DNA fragments are transferred to a membrane (Southern Blotting).
- The single-stranded gene probe is applied to the membrane; if a fragment(s) contains the gene, the probe will bind to it.
- If a radioactive probe is being used, a photographic film is applied to the membrane to detect the radioactive signals.
- Development of the film will disclose whether or not the individual being tested possesses the mutated gene being sought; for example, a mutated gene may be the result of a deletion of a portion of the gene; therefore, the number or sizes of the DNA fragments (produced by treatment with restriction enzymes) will be noticeably reduced.

For reasons of safety and cost, many laboratories prefer to use non-radioactive labels for probes (see Chapter 10).

Restriction Fragment Length Polymorphism (RLFP) Analysis for a Particular Gene Defect

Deletion or alteration of a restriction site results in an RFLP. In higher organisms, the individual fragments resulting from restriction endonuclease digestion of their complex, large genomes are not seen as separate bands after gel electrophoresis because there are too many DNA fragments.

The following approach is used to detect the RFLP:

• The restriction digest is subjected to electrophoresis.
• The separated fragments are transferred to a nitrocellulose or nylon membrane (Southern Blotting).
• Analysis by hybridization is carried out using a labeled probe that is complementary to the region of interest.
• The probe will only hybridize to the restriction fragments in the relevant area.
• The sizes of the fragments detected by the probe will be apparent by comparison to molecular weight standards.
• The fragment sizes are compared with the sizes obtained for the unmutated gene.
• If an RFLP is present, it will be clearly apparent as a shift in the number or size of one or more DNA fragments (see Figure 5.1).

RFLP Linkage Analysis

This is a more useful method for screening for a defective gene than the one just described. It requires that the RFLP is present in the vicinity of the defective gene. If this RFLP is close enough, it will be inherited with the defective gene as it is not likely that a recombination event will separate the gene from the RFLP during meiosis.

It has been useful to study families with a high incidence of the disorder (i.e., the defective gene) by comparing the particular RFLP to those families without the defect. This approach has been used to find a marker gene or sequence for the diagnosis of Huntington disease, cystic fibrosis, Duchenne muscular dystropy, etc. The presence of the RFLP will indicate the defective gene.

With some genetic diseases the chances that the RFLP and the defective gene will ever be separated is very small and thus the presence of the RFLP can be considered diagnostic for the defective gene. Most RFLPs of this kind are known because the neighboring genes have been cloned and sequenced.

Direct Identification of the Mutated Gene

See Allele Specific Oligonucleotide Analysis below under Breast Cancer.

GENETIC SCREENING/TESTING

Genetic screening and testing procedures are outlined as follows:

Prenatal Screening

See Amniocentesis, Chorionic Villus Sampling and Multiple Spectral Imaging below.

Newborn Screening

This is the most widespread type of genetic testing. The tests involve examination for mutant genes associated with the more common inherited disorders as well as tests for errors of metabolism, e.g., in phenylketonuria.

Carrier Testing

This is used to help married couples learn whether or not they carry a recessive allele for various inherited diseases, e.g., cystic fibrosis, sickle cell anemia, or Tay-Sachs disease.

Predictive Gene Testing

This involves tests that identify individuals who are at risk of getting a disease as a result of a genetic disorder. At present there are tests for at least 24 such diseases. Predictive gene tests are particularly for disorders that "run in families" as a result of the inheritance of a faulty gene. Predictive gene tests are available for cystic fibrosis, Tay Sachs disease, Huntington disease, some forms of Alzheimer's disease, predisposition to amyotrophic lateral sclerosis (Lou Gehrig's disease), and inherited tendencies toward development of some common cancers.

Amniocentesis

This is the most common prenatal screening procedure. The test is conducted from the 14th to 16th week of pregnancy. A hollow needle is inserted through the mother's abdominal wall into the uterus then into the amniotic sac surrounding the fetus. The amniotic fluid obtained contains some fetal cells. The cells are cultured for 10 to 12 days to obtain sufficient numbers for chromosomal and biochemical analysis. PCR (polymerase chain reaction) can be used to amplify the region containing the gene.

Amniocentesis results in miscarriages in 0.3 to 3 percent of cases. Simultaneous ultrasound imaging has helped lower the rate.

Chorionic Villus Sampling (CVS)

This prenatal test can be carried out as early as the 8th and 9th week of pregnancy. Chorionic cells are obtained by introducing a catheter into the uterus until it touches the chorionic villi. The cells collected are used for genetic and biochemical analysis. Unfortunately, CVS triggers miscarriages at about the same percentage as does amniocentesis.

Multiple Spectral Imaging (Spectral Karyotyping)

In this recently developed screening procedure, fluorescent dyes are attached to different chromosomes and as a result, each pair of chromosomes has a different color, e.g., chromosome 1 is colored yellow, 2 is red, 3 is gray, and so on. The differential fluorescence is made possible by the *in-situ* (in place) hybridization with probes tagged with different fluorescent dyes.

The procedure is particularly useful in detecting chromosomal defects. For example, the extra chromosome causing Down syndrome is readily recognized, and a piece of one chromosome breaking off and reattaching to another chromosome (translocation), as occurs in some cancers, is easily detected.

Spectral imaging can spot duplications or deletions of DNA within a single chromosome that are characteristic of lung, breast, and other cancers. The so-called "marker chromosomes" of cancer cells that are made up of pieces of chromosomes, e.g., 3, 8, and 13 are readily recognized. Klinefelter's syndrome, in which an individual has two X chromosomes and one Y, is likewise easily diagnosed.

Certain genes sometimes present on particular chromosomes make tumors more resistant to standard cancer treatment. Recognition of these genes by spectral imaging can be helpful in redirecting therapy.

The potential of this procedure is indeed great. Eventually spectral karyotyping may be done by machines thus greatly speeding up the process.

General

Because of the danger of miscarriage, current prenatal tests for genetic disorders are mostly restricted to women 35 years or older. They are at greater risk of having a child with Down syndrome.

There is an urgent need for a prenatal test without the disadvantages of amniocentesis and CVS. Much research is being directed to developing a test that would only require a blood sample from the mother. The most promising procedure to date would appear to be examination of fetal red blood cells obtained from the mother's blood.

The pros and cons of genetic screening are discussed in Chapter 12. Screening for such disorders as Down syndrome, Tay-Sachs disease, cystic fibrosis, and phenylketonuria would seem to be of obvious value. In the latter two conditions, diagnosis makes possible the initiation of life-saving measures. The extent of current prenatal genetic screening varies among states in the U.S., and in countries able to provide screening.

SOME HUMAN GENETIC DISORDERS

When considering those genetic disorders due to recessive genes, it is helpful to recall the laws of Mendelian genetics. The offspring of two carriers has:

- A 25 percent chance of developing the disorder, i.e., of inheriting two aberrant genes.
- A 50 percent chance of being a carrier, i.e., acquiring one of the aberrant genes.

• A 25 percent chance of not inheriting the aberrant gene.

In the case of autosomal dominant conditions such as Huntington disease, the affected individual risks transmitting the dominant mutant to half of his offspring.

Bilateral Retinoblastoma

A rarely occurring malignant neoplasm of retinal cells usually in children less than three years old. This dominant gene (NF) is located on the long arm of chromosome 13. The NF gene resembles Ras-GAP gene which interacts with the well-known oncogene called ras.

Color-Blindness

The genes responsible for color-blindness (red blindness and green blindness) occupy different loci on the X chromosome (sex-linked). There is a reduction or absence of one of the visual pigments; blue blindness is very rare.

Cri du Chat (Cry of the cat)

Manifestations are microencephaly, mental deficiency and a characteristic plaintive, cat-like cry. This rare disorder affects infants and is due to deletion of the short arm of chromosome 5.

Cystic Fibrosis (CF)

The most common hereditary disorder of Caucasian children; about 1 in 2300 births. One in every 25 Americans is a carrier of the CF gene. CF is due to a recessive mutant gene on chromosome 7. The CF gene is about 250 kb long; its 24 exons encode a protein with 1,480 amino acids.

This devastating disease results from the malfunction of mucus producing cells of the body. Most patients with CF die from respiratory failure. Air passages in the lungs become clogged with mucus. Antibiotic treatment is of value in coping with infections and thus greatly prolonging life.

Down Syndrome

It is due to three copies of chromosome 21 (trisomy 21) and characterized by severe mental deficiency. About one of every 200 infants is affected. The frequency increases with advancing maternal age.

Duchenne Muscular Dystrophy

It is due to a recessive gene on the distal portion of the short arm of the X chromosome. This sex-linked disorder affects about 1 in 3500 boys and involves the protein dystrophin. The disorder is characterized by progressive muscular weakness.

Familial Adenomatosis Polyposis (FAP)

This is a name for two genetically dominant disorders: familial polyposis coli, and Gardner's syndrome. Both conditions are characterized by the growth of numerous polyps in the colon of persons usually under 30 years of age. If they are not removed, there is a high risk of cancer developing. FAP is a polygenic disorder, i.e., there is an interaction of several genes.

Fragile-X Syndrome

It is thus named because the X chromosome is changed so that it tends to break when prepared for observation. The defect is the result of a long repeat of the triplet CGG. Next to Down syndrome, it is the most common cause of mental retardation currently identified.

Galactosemia

This disorder is due to an autosomal recessive gene on the short arm of chromosome 9. There is a deficiency of the enzyme for utilization of galactose resulting in the enlargement of liver and spleen. Removal of galactose from the diet is effective in controlling the disorder.

Hemophilia A

This is a sex-linked (gene on X chromosome) disease in which blood clotting is impaired (clotting factor VIII) leading to the risk of uncontrolled bleeding. It is transmitted through the female and usually affects males; it may be linked to color-blindness.

Huntington Disease

This disorder is due to a single dominant gene occupying the tip of chromosome 4. The gene is referred to as a "stuttering gene" because it has many more triplet (CAG) repeats than is normal. The protein huntingtin is defective resulting in degeneration of the nervous system leading to chorea with death usually in early middle age.

Several genetic disorders are due to "stuttering genes," e.g., spinobulbar muscular dystrophy, fragile X syndrome, and myotonic dystrophy.

Klinefelter Syndrome

This disorder results in sterile males with small testes and no sperm. The individual has two X chromosomes and a Y chromosome.

Lesch-Nyhan Syndrome

An X-linked recessive disease affecting boys that results in mental retardation and compulsive self-mutilation. It is caused by a defect in the gene that produces a particular enzyme involved in purine metabolism. The result is an excessive production of uric acid. The mutation involved occurs spontaneously in one in 10,000 boys.

Marfan Syndrome

This disorder, which President Lincoln had, involves connective tissue and is inherited as an autosomal dominant. It is characterized by very long fingers ("spider fingers") and toes. The prevalence is 1/20,000.

Phenylketonuria (PKU)

A recessive disorder involving two copies of the mutant PKU gene. The enzyme for metabolizing the amino acid phenylalanine is lacking and phenylketonuria results. About one in 15,000 babies are affected. The toxicity produced by excess phenylalanine causes brain damage leading to mental retardation. When the deficiency is recognized, the individual is placed on a special low protein diet and damage is averted.

Sickle-cell Anemia

A severe hereditary blood disorder, mainly of persons of African ancestry. It is due to a defect in a single gene. Those inheriting two copies of the mutant gene develop the disorder. Inheritance of one mutant gene and one normal gene, i.e., the sickle-cell trait, does not result in the disease but provides protection against malaria.

The disease is due to a single amino acid substitution in the beta-chain of the normal hemoglobin molecule. Red blood cells have a sickle shape and they block capillaries with subsequent anemia. 0.2 percent of African American babies in the U.S. suffer from sickle-cell anemia

Tay-Sach Disease

This lethal disorder is characterized by degeneration of the nervous system. It is most common in people of Jewish ancestry and in French Canadians. The recessive gene results in the lack of the essential enzyme hexosamidase A leading to the accumulation of its substrate.

Turner syndrome

Affected females have only one X chromosome. The ovaries are rudimentary or missing and those affected are sterile.

CANCER

Cancer cells, which arise from normal cells, are permanently changed and are not under the normal controls which regulate the division of normal cells. These altered cells multiply more rapidly than normal cells and they may eventually spread or metastasize by the blood and lymphatic systems to other tissues which they invade and destroy.

Proteins called growth factors are associated with normal cells. They speed up cell division or slow it down (tumor suppressors). The process of cell division can go wrong in the following ways:

- If there is a defective gene, the cell may produce proteins that initiate cell division in the absence of growth factors.
- The DNA may be damaged to the extent that it can't produce the proteins that shut the cycle down (tumor suppressors).
- The control system can fail if one allele mutates so that it results in the production of too much of the protein that initiates cell division.

Oncogenes (genes responsible for transforming normal cells into cancer cells) are examples of this failure. Some oncogenes are mutant versions of normal cellular genes.

Both alleles at a locus can be mutated to the point that an essential protein, such as a tumor suppressing factor, is not produced. This would result in an abnormal proliferation of cells.

If you inherit a defective gene from one of your parents, you are all right if you have one functioning allele. But if the functioning gene is damaged by a mutation, you then have an enhanced susceptibility.

According to the "clonal evolution" or "multiple-hit theory" of cancer causation, the control systems can manage with one or two mutations. However, if there are several "hits" involving various genes, cancer may develop. "Hits" refer to damage to DNA resulting from radiation, chemical carcinogens, radiation, free radicals, and other oxidants.

Colon cancer has been investigated thoroughly. The following steps in its development are recognized:

- Loss of the tumor suppressing gene APC on chromosome 5 resulting in greater proliferation of epithelium (cells lining the intestinal tract) with development of early adenoma (benign tumor).

- Activation of the growth promoting gene K-ras resulting in intermediate adenoma.
- Loss of the tumor suppressing gene DCC on chromosome 17 resulting in late adenoma.
- Loss of the tumor suppressing gene p53 resulting in carcinoma (malignant tumor) leading if unchecked to metastasis (spread).

Because five genes must mutate for colon cancer to develop, the disease is usually seen in older people.

Other cancers are thought to develop by a somewhat similar sequence of events. Knowledge of these events for each particular cancer should contribute greatly to their prevention and treatment.

Some cancers such as retinoblastoma (RB), a cancer of the retina of the eye of young children, occurs when a child has both mutant copies of the single gene RB.

There are a number of oncogenic viruses that can transform the cells they infect with the result that the infected cells proliferate in an uncontrolled way. Many of the oncogenic viruses are retroviruses, i.e., RNA viruses that use reverse transcriptase in their life cycle.

Gene p53

The role of the gene *p53* in colon cancer has been mentioned above. There is now evidence that alterations in this gene are involved in more than 50 kinds of cancer. Intensive research in many laboratories is currently directed to elucidating the role of gene p53 in the development of particular cancers.

The normal gene *p53*, which consists of 2,362 base-pairs, produces a protein that helps the cell cope with damage to its DNA from ionizing radiation or chemical carcinogens. Genes located nearby *p53* are triggered by *p53* to produce proteins that inhibit uncontrolled cell proliferation. When *p53* fails, it can be responsible for an estimated 60 percent of human cancers, including those of the lung, breast, colon, prostate, liver, bladder, cervix, and skin.

Failure of gene *p53* protein is often due to a point mutation. The gene may be altered so that it cannot produce its tumor suppressing protein. Most *p53* mutations are not inherited but result from carcinogens or from an error in copying DNA.

Benzopyrene in cigarette smoke; aflatoxin (from spoiled peanuts, corn, etc.); hepatitis-B virus; human papilloma virus and ultraviolet light may be responsible for changes in *p53* that lead, respectively, to cancer of the lung, liver, cervix, and skin.

Knowledge of some of the mechanisms whereby gene *p53* contributes to the development of cancer are already contributing to more effective treatment of the disease by radiation and chemotherapy. Investigators are particularly optimistic about future prospects of this area of cancer research.

Breast Cancer

The two genes *BRCA1* and *BRCA2* help to suppress the unrestrained proliferation of cells that can lead to cancerous breast tumors. The failure of either gene due to mutation can lead to breast cancer. A defective *BRCA1* gene has also been implicated as a cause of ovarian cancer. The association of mutant *BRCA* genes with the occurrence of breast cancer is complex and genetic tests have thus far been poorer at predicting cancer than earlier studies had suggested.

The procedures summarized below are being used to screen for defective *BRCA* genes:

1. *Sequencing the Relevant Genes to Determine Defects.* This procedure is time-consuming and costly. The methods that follow are quicker and less expensive.

2. *Multiplex Polymerase Chain Reaction (MPCR).* This procedure employs primers to specific loci to locate a particular mutation due to a deletion. The enzyme DNA polymerase is used to greatly multiply the defective DNA (see PCR Chapter 6). Gel electrophoresis is used to determine whether or not the amplified loci within the DNA contains a deletion based on the sizes of the fragments.

3. *Allele Specific Oligonucleotide Analysis (ASO).* Artificial oligonucleotides (oligos) are designed to pair precisely with a particular defective stretch of DNA. The oligos will adhere to the single-stranded DNA possessing the mutated gene but not to the normal unmutated genes. The oligos are tagged with a fluorescent dye so the location of the mutation can be determined.

4. *Confirmation Sensitive Gel Electrophoresis (CSGE).* The sample DNA, reference DNA possessing normal genes, and DNA with known mutations are treated with appropriate restriction enzymes and the fragments generated are compared using gel electrophoresis. The

results will determine the likelihood that the sample DNA contains one or both of the breast cancer genes.

5. *Protein Truncation Test (PTT)*. PTT looks at the proteins produced by defective genes. Because the mutant BRCA genes are truncated (e.g., contain a premature stop codon), they produce proteins that are small. The latter are identified by a gel electrophoresis procedure devised to separate proteins based on size.

Chapter 10

DNA PROBES FOR DIAGNOSTIC PURPOSES

DNA probes have been referred to in several earlier chapters. In Chapter 9, we discussed the use of probes in detecting genes responsible for various heritable disorders. Here we are mainly concerned with their use in the laboratory diagnosis of infectious diseases.

DNA PROBES FOR DIAGNOSTIC PURPOSES

A probe is a fragment of DNA or RNA (usually about 50 to 100 nucleotides long) that is labeled radioactively, or with a fluorescent dye or other compound (see below). The labeled fragment seeks out by hybridization, its complementary nucleic acid in, e.g., a microorganism, and thus makes possible the latter's identification. Probe-based diagnostic procedures are sensitive and rapid.

Probes must have a high degree of specificity. As is mentioned in Chapter 8, they can be chemically synthesized and represent isolated regions of a gene, or cloned intact genes. In the example outlined below, the extracted DNA is cut with a restriction endonuclease followed by cloning in a plasmid.

In general, the laboratory procedure involves the following steps:
- The single-stranded DNA from the clinical specimen is bound to a membrane support.
- The single-stranded labeled DNA probe is added to the membrane (target DNA) to allow base-pairing.
- Is washed to remove the unbound labeled DNA.
- Observe whether or not a signal is emitted by the labeled probe, e.g., radioactivity or fluorescence, as the result of base pairing to target DNA.

87

Probes have been developed that hybridize target DNA in tissue samples, blood, feces, urine, and respiratory secretions.

As was mentioned in Chapter 4, if the number of the target sequence is very low in clinical samples from some infections, the polymerase chain reaction can be used to amplify the sequence.

NONRADIOACTIVE HYBRIDIZATION PROCEDURES

For reasons of safety and cost (less costly equipment), nonradioactive hybridization procedures are preferred by many diagnostic laboratories to those methods using a radioactive label.

One nonradioactive procedure has the following steps:
• A biotin labeled DNA probe is hybridized to the target DNA of a clinical sample.
• Avidin, which has a high affinity for biotin, is added.
• Biotin labeled with an enzyme such as peroxidase is added.
• To determine if hybridization has taken place, i.e., the avidin has combined with the biotin of the probe, an appropriate substrate is added.
• If enzyme is present, i.e., biotin–avidin–biotin have combined, the chromogenic substrate is reacted upon resulting in a color change that can be readily detected.

If the chemiluminescent substrate luminol is used with the enzyme horseradish peroxidase, the luminescent response (light) obtained as a result of the enzymatic degradation can be detected with a photographic film.

A PROBE FOR THE DETECTION OF *SALMONELLA*

A probe for the bacterium *Salmonella* is prepared by extracting the DNA from it. This DNA is cleaved into fragments by a restriction enzyme; a specific fragment is selected and purified for use as a probe. It must be able to hybridize with the DNA of all *Salmonella* strains but not with closely related enteric (intestinal) bacteria.

The selected fragment is cloned into a plasmid which is introduced into *E. coli* by transformation (see Figure 10.1). The growth of *E. coli* thus produces hundreds of the specific *Salmonella* DNA fragments (in plasmids). The plasmids are purified, the fragments are extracted and tagged with a fluorescent dye or other label.

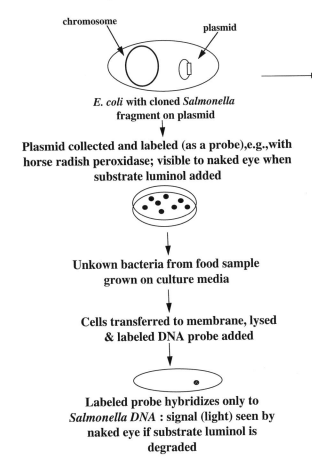

Figure 10.1. Steps involved in the preparation and use of a DNA probe for identification of *Salmonella* in a food sample.

The unknown bacteria are collected on a nitrocellulose filter; cells are lysed to release the DNA. The labeled *Salmonella* DNA probe is applied to the filter containing the unknown DNA. If the *Salmonella* probe hybridizes with a portion of the unknown DNA, then the latter is assumed to be from a *Salmonella* bacterium. The label makes possible the detection of hybrids formed between the probe and the unknown DNA.

Chapter 11

VACCINES AND IMMUNOLOGICAL PROCEDURES

There are a number of drawbacks to traditional inactivated and live attenuated vaccines including low immunogenicity and untoward side effects. Some of the latter are due to endotoxin (complex lipopolysaccharide molecules composing the cell wall of some bacteria) and may produce shock.

RECOMBINANT VACCINES

Recombinant DNA technology is being used in various ways to develop safer and improved vaccines.

Two principal approaches are being used in the development of recombinant or subunit vaccines:

Use of Expression Vectors*

Genes for two foot-and-mouth disease virus proteins have been cloned in a nonpathogenic *E. coli* strain. Large amounts of the two protein antigen molecules were produced, but only one has proved effective as a vaccine.

The core protein of the hepatitis B virus has been successfully produced in recombinant *E. coli*. Although not used as a vaccine, the purified protein is used in laboratory diagnosis.

---

* Defined at the end of this chapter.

Another example of a subunit vaccine is one against enteropathogenic *E. coli.* The antigen, the beta subunit of the enterotoxin of *E. coli,*
was cloned and inserted into a nonpathogenic strain of *E. coli.*

Not all recombinant subunit vaccines have been effective.

Purification may be difficult and pure proteins are not always good
immunogens (stimulators of immunity). However, the ability to produce an immunogenic protein from a disease-producing microbe via
a nonpathogenic recombinant strain provides a large measure of safety.

Use of Virus Carriers

Recombinant vaccinia viruses have been used as live vaccines
against several diseases. As a carrier, vaccinia virus has a number of
advantages:
  • It replicates in the cytoplasm rather than in the nucleus.
  • Considerable DNA can be replaced without interfering with repli-
    cation.
  • It produces a very low grade infection.
  • It is a large virus and can accommodate two or three dozen foreign
    genes.
  • It is a "laboratory" virus that does not occur in nature.
  • It can express high levels of new antigen.
Viral genes have been inserted into a region of the vaccinia genome
called thymidine kinase (TK) gene.

A number of foreign genes have been expressed by recombinant
vaccinia viruses including:
  • Hepatitis B major surface antigen
  • Rabies virus G protein
  • Vesicular stomatitis virus proteins
  • A number of other viral proteins
A disadvantage of vaccinia and other viral carriers is that they elicit an immune response that may interfere with subsequent inoculation
of the same virus carrier. Another disadvantage is that individuals with
an impaired immune system may develop the serious complication,
postvaccine encephalitis.

Adenoviruses and herpesviruses have also been used as carriers for
expression of foreign antigens.

LIVE MODIFIED VACCINES

Some strains of pathogenic organisms have been engineered so that their virulence genes have been deleted or modified. A live cholera vaccine has been developed in which sections of the gene encoding the "toxin" sequence were deleted resulting in no enterotoxin production and thus no disease. The genes for natural immunogens were retained and thus the nonpathogenic strain induces protective immunity against the pathogenic strain.

ANTI-IDIOTYPE VACCINES

Anti-idiotype antibodies are antibodies produced against antibodies that bind to a particular antigen. Some of these anti-antibodies have the antigenic characteristics of the original antigen. Anti-idiotype antibody has thus far only been used experimentally to immunize humans and animals.

DNA VACCINES

This approach to vaccination, also called Genetic Immunization, is still in the early stage of development.

Generally speaking, infectious diseases can be divided into two major types. Those caused by intracellular microorganisms and characterized by a cellular or cell-mediated immune response. Examples are tuberculosis, brucellosis, listeriosis, and many fungal and viral diseases. The other category of infections are those caused by mainly extracellular microorganisms which elicit a humoral or antibody-mediated response. Many bacterial pathogens, including streptococci and staphylococci, are in this category.

DNA vaccines are being used, thus far experimentally, to immunize against the first kind of infections, i.e., those caused by intracellular microbes. Traditional vaccines that elicit mainly humoral immunity have not been very effective in preventing these "intracellular" diseases. DNA vaccines elicit both cellular and humoral immunity.

Intracellular pathogens such as the tubercle bacillus and viruses are shielded from antibodies (humoral immunity). Bits of protein from the intracellular pathogen are presented on the surface of the host cell. These proteins signal the cellular immune system that this cell should be destroyed. Killing is accomplished by killer T-cells and spread of infection is prevented. This immune response is an example of a cellular (cellular immunity) immune response as opposed to a humoral response (antibody).

The following steps have been involved in the development of a DNA vaccine:

1. A copy of the gene encoding the pathogen's protein is obtained, e.g., in the case of a particular virus, the capsid or coat protein.

2. The gene is inserted in a plasmid and many copies of the gene are obtained by cultivation of the bacteria containing the plasmid.

3. The plasmids containing the gene for the pathogen's protein are harvested and purified.

4. The plasmids, constituting the vaccine, are injected into an experimental animal susceptible to the viral disease being studied.

5. The plasmids find their way into the animal cells and the virus gene is transcribed into the protein.

6. The pathogen's protein adhering to the animal cell's surface activates the cellular immune system which enables the animal to resist infection.

Some of the advantages of DNA vaccines are the following:

• They are economical to produce.

• Unlike live vaccines, they cannot cause infection.

• Several plasmids coding for several proteins could, theoretically, be administered at one time.

• Because DNA is particularly resistant to deterioration, DNA vaccines are more stable than traditional vaccines. Refrigeration is not always available in developing countries.

• They elicit both cellular and humoral immunity.

Clinical trials are planned or are already underway to use DNA vaccines to prevent hepatitis B, HIV (human immunodeficiency virus), influenza, and malaria. Investigators propose that DNA vaccines may be useful in dealing with various autoimmune diseases, cancer, and allergies.

IMMUNOLOGICAL PROCEDURES

Many immunologically-based detection procedures are employed in diagnostic laboratories. The ELISA (enzyme-linked immunosorbent assay) is a particularly sensitive immunological procedure with many applications. Monoclonal antibodies (a single type of antibody that is directed to a specific antigenic determinant or epitope) increase the specificity of ELISA diagnostic tests; however, they are expensive to produce. To reduce the cost of monoclonal antibodies, a cloning strategy for expressing portions of genes encoding antibodies has been developed for producing antibody fragments in *E. coli*. These single-chain antibodies produced in *E. coli* have many diagnostic applications.

Chapter 12

MICROBIAL SYNTHESIS AND
COMMERCIAL PRODUCTS

PRODUCTION OF COMMERCIAL PRODUCTS

The production of recombinant proteins is a major activity of recombinant DNA technology. After an efficient expression system has been developed, it is not usually difficult to clone and express the protein of interest (see Expression Vector below).

The expressed protein may be the final product or a catalyst for various chemical reactions. Such catalytic activity is taken advantage of in the production of many compounds including, dyes, amino acids, antibiotics, and vitamins. The production of such compounds can be efficient and relatively inexpensive. Several different strategies are employed to isolate either the cDNAs or genes for animal or human proteins:

- In one approach the target protein is isolated and the amino acid sequence or a portion of it is determined. This makes it possible to produce a DNA coding sequence. The appropriate oligonucleotide is synthesized and then used as a DNA hybridization probe to isolate the cDNA or gene from either a cDNA or a genomic library.
- Another approach is to prepare a specific antibody against the protein and use it to screen a gene expression library.
- For some proteins like insulin, which is synthesized in a single tissue, a cDNA library from mRNA of the tissue will be enriched or added to identify/facilitate isolation of the target DNA sequence.

Rather complex procedures may be involved in isolating cDNAs, as for example, with the three interferons. The interferon cDNAs are subcloned into *E. coli* where they are expressed at high levels.

More than 50 human and animal proteins have been produced by recombinant DNA methods. Some of the better known ones and their uses are the following:

- Human growth hormone: to treat growth stature deficiency
- Adrenocorticotrophic hormone: to treat rheumatic disease
- Bovine growth hormone: increases bovine milk production
- Erythropoetin: to treat anemia
- Insulin: to treat diabetes
- Interleukins: to treat cancer and immune disorders
- Serum albumin: constituent of blood plasma
- Tumor necrosis factor: an anticancer agent

EXPLANATORY

Expression Vector

This is a cloning vector made in such a way that after the insertion of a DNA molecule, its coding sequence is transcribed and the messenger RNA effectively translated into protein.

Considerable amounts of the protein encoded by foreign DNA can be produced in cells by choosing appropriate expression vectors. The cloned gene is under control of a promoter sequence for the initiation of transcription, etc., that is recognized by the host cell. This often allows for the expression of eukaryotic cDNA in prokaryotic cells.

Chapter 13

GENETIC ENGINEERING OF ANIMALS

TRANSGENIC ANIMALS

The idea in producing transgenic animals is to add single, functional genes into the chromosomes of the animal; this additional gene(s) would not occur if normal mating took place. The approach used can be summarized as follows:
- A cloned gene is injected into the nucleus of a fertilized egg.
- These fertilized eggs are implanted into the female.
- Some of the progeny of the implanted eggs will carry the cloned gene in all their cells.
- Those animals with the cloned gene in their germ line cells are bred to initiate new genetic lines.

What happens to the transgenic animal will depend upon the product encoded by the injected gene, e.g., the product may result in greater growth and feed efficiency.

An animal is said to be transgenic when its genetic make-up has been altered by the addition of exogenous (foreign) DNA. The DNA involved is referred to as a transgene and the whole process is termed transgenesis or transgenic technology. Most of the pioneering work on transgenesis has been done with the laboratory mouse.

The following three strategies have been used:

Virus Vectors

Retroviruses have been used to introduce transgenes into the genome of mammalian cells. These viruses can only accommodate small pieces of DNA, about 8 kb. Another major drawback is the introduction of an infectious virus which may insert itself into a chromosome and cause a mutation, e.g., cancer even if rendered defective.

Microinjection

Because of the disadvantages just referred to, DNA microinjection is preferred. It involves injecting the transgene into the male pronucleus with a fine, microscopic needle while the egg is held in place under a dissecting microscope. In the case of mice, the eggs are implanted into a foster mother.

The transgenic offspring can be identified by assaying tissue, e.g., blood by Southern blot hybridization or by the polymerase chain reaction using probes specific for the transgene. Identifying the gene in offspring will indicate whether or not the transgene is in the germ line.

Embryonic Cell Method

Cells from the blastocyst stage of embryo development in the mouse can proliferate and retain their capacity to differentiate into all other cell types after being introduced into another blastocyte embryo. These cells are called pluripotent embryonic stem (ES) cells.

A transgene can be integrated within a dispensable DNA sequence in the genome of the pluripotent embryonic stem cells. These cells can then be cultured, inserted into the blastocystic stage embryo and used to generate transgenic animals.

ES cells, like those of the mouse, have not yet been found in birds, sheep, goats, pigs, chickens, and cattle.

CLONING GENETICALLY INDISTINGUISHABLE ANIMALS

This very recent procedure, outlined below, has thus far only been applied to sheep.
- Cells are isolated from sheep embryos and cultivated to produce large numbers of cells.
- Unfertilized ova are taken from ewes; the genes in the form of nuclei are extracted.
- The ova, less their genes, are fused with the cultivated embryonic cells which provide identical sets of genes.

- The ova with the "new" genes are placed in ewes to grow and develop.
- The result is the production of identical lambs.

CLONING AN ADULT SHEEP

The cloning of an adult sheep was achieved in a Scottish laboratory in 1997. This landmark achievement, thought by many to have been impossible, had a profound effect throughout society because it raised the unsettling possibility of cloning human beings. Shortly after the cloning was announced, steps were hurriedly taken in the United States to prohibit the use of federal funds for research on cloning human beings. A number of countries including the U.S.A. have laws prohibiting human cloning.

The steps involved in the cloning of the adult sheep are outlined below. That success did not come easily is evident from the fact that of 277 tries only 29 embryos survived longer than six days. All died before birth except the lamb named Dolly.

- Cells were taken from the udder of an adult Finn Dorset ewe and placed in a culture medium with very low concentrations of nutrients; cells stopped dividing and switched off their active genes.
- An unfertilized ovum was taken from a Scottish Blackface ewe; the nucleus (containing the DNA) was removed from the ovum leaving an egg cell containing what is necessary to stimulate a nucleus to produce an embryo.
- A nucleus was removed from the udder cell of a Poll Dorset ewe and placed next to the enucleated ovum (from the Blackface ewe) and subjected to the pulse of an electric current which caused them to fuse and start cell division, i.e., embryonic development.
- About six days later the developing embryo was implanted in another ewe.
- At the end of the gestation period (145 days), the ewe gave birth to a Poll Dorset lamb which was found to be genetically identical to the original nucleus donor, the Poll Dorset ewe; the lamb was named Dolly.

Among the possible benefits of cloning adult mammals are: propagation of endangered species, producing (in animals) temporary replacement organs for transplant patients, cloning of "champion" livestock, and replication of genetically engineered animals useful in research and in the production of bio-pharmaceutical drugs.

Shortly after the cloning of Dolly, five more female lambs were cloned at the same laboratory in Scotland. These lambs, in contrast to Dolly, developed from cells obtained from an ovine fetus. One of these lambs, Polly, carried a human gene that was inserted when it was a single-celled embryo. The human gene was inserted so that a medically valuable protein would be secreted in the sheep's milk.

It is planned to use this cloning technique to produce transgenic cattle and swine in addition to sheep for the production of various useful proteins.

Chapter 14

GENETIC ENGINEERING OF PLANTS

TRANSGENIC PLANTS

Gene modification of plants has resulted in a number of benefits and there is a potential for many more. Two vector systems have been used: the Ti plasmid of the bacterium *Agrobacterium tumifaciens* and some plant viruses. These systems and direct gene transfer are summarized below. As far as is known, plasmids do not occur naturally in plants.

Ti Plasmid

A. tumifaciens is a soil bacterium that causes crown gall disease in a number of dicotyledon plants. Crown gall (a tumor) occurs as a result of a wound to the stem of the plant. The crown gall bacterium gains entrance via the wound and infects the plant tissue producing a cancerous proliferation of plant cells (crown gall).

The proliferation of cells is associated with the presence of the Ti (tumor inducing) plasmid within the bacterial cell. It is a large plasmid, >200 kb, that has numerous genes. After infection of plant cells, part of the Ti plasmid is integrated into the plant chromosomal DNA. The integrated segment called T-DNA ranges from 15 to 30 kb in size; it is passed to daughter cells as an integral part of the chromosome.

By deleting the tumor inducing genes of the Ti plasmid, and subtituting a gene of interest the engineered plasmid will still integrate into the plant genome and express the foreign gene. A variety of transgenic plants have been developed, including those which are resistant to certain insects and herbicides as well as ones which exhibit delayed ripening characteristics to facilitate shipment over long distances.

Transgenic potatoes expressing a mutated bacterial toxin when fed to mice induce specific antitoxin antibodies and open the possibility of using plants as food based vaccines.

Cloning with Plant Viruses

Most of the viruses infecting plants have RNA genomes. RNA viruses are less useful than DNA viruses as potential cloning vectors

Two classes of viruses are known to infect plants: the caulimoviruses (includes the cauliflower mosaic virus, CaMV) and the geminiviruses. Although CaMV has been used as a vector it has the disadvantage of limited capacity for carrying inserted DNA and a limited host range.

The geminiviruses also have a limited host range, but it includes the important plants, maize and wheat. Geminiviruses have small genomes and infection cannot be accomplished by convenient surface inoculation. Two disadvantages of the geminiviruses as vectors are: instability of the viral genomes after infection and the potential of these viruses to cause crop diseases. Additional research may overcome these disadvantages.

Direct Gene Transfer

Some bacterial plasmids can be introduced into plant protoplasts or shot in using a gene gun and although they won't replicate on their own, integration into the plant chromosome by recombination is possible. If the plasmid introduced into the protoplast carries a gene, the latter may be integrated into the chromosomal DNA. If maintained, it may pass to progeny. One problem has been the regeneration of a plant from protoplasts. The technique has the advantage of being applicable to all plants and it seems likely, with more research, to be the preferred method.

Chapter 15

GENE THERAPY

G ene therapy is a young field with what many consider a great potential to ultimately correct or alleviate certain human genetic disorders. Much of the research in human gene therapy, which is currently very intensive, is concentrated on correcting defects in somatic cells (nonsex cells). The number of successes achieved with this therapy to date has been limited in both animal and human trials.

The main strategies of gene therapy are summarized below.

EX VIVO GENE THERAPY

This is the type of manipulation most commonly called gene therapy. It involves the following steps:
- Remove cells from patient (e.g., cells lining the lungs in patient with cystic fibrosis).
- Place in dish of nutrient medium for growth and multiplication
- Introduce to the cells (transform) the normal gene via, e.g., a retrovirus (altered so that it is not harmful).
- Reintroduce the cells into the patient from which they were obtained.
- Ideally, cells with the transplanted gene produce the desired protein.
- As new genes are not passed to future generations of cells periodic transfusions of the transformed cells will be necessary to correct the defect.

If a retrovirus is used as a vector, the gene of interest is inserted as an RNA copy into the viral RNA genome. The virus injects its RNA along with the RNA gene copy into the cell and DNA is made from the RNA by the enzyme reverse transcriptase. The DNA becomes part of the chromosome in the nucleus and the cell can then produce the proteins that are encoded by the introduced genetic material.

Difficulties with this kind of therapy include the problem of gene expression in the target cell, determining what are the target cells in some disorders, and how to get the corrective genes to the affected cells. For example, it is difficult to get viral vectors into bone marrow stem cells. The latter divide infrequently; their frequency of division facilitates entrance of viral vectors. *Ex Vivo* Gene Therapy can also be limited by the difficulty of *in vitro* cultivation of some cells, e.g., those of the brain.

A notable success of the *ex vivo* approach was the treatment of a child with severe combined deficiency disorder (humoral and cellular immune deficiency). White cells were recovered from the patient and exposed to viruses carrying the gene for the missing vital enzyme. The infected white blood cells were transfused into the child and provision of the normal gene resulted in production of the vital enzyme. Repeated treatments averted the damaging effects of this disorder.

Viruses other than retroviruses are being used as vectors in both *ex vivo* and *in vivo* therapy. These include adenoviruses, adeno-associated viruses, alphaviruses, herpesviruses, and poxviruses. Various viral vectors have been discussed in Chapters 11, 13, and 14. Each group has its own advantages and drawbacks, some of which were discussed in these chapters. Some of the drawbacks are as follows:
   • Some viruses are more susceptible to attack by the immune system than others.
   • Small viruses have limited capacity for maintaining foreign genes.
   • Some genes can only function for a short time due to failure of viral integration.
   • Some viruses only infect dividing cells.
   • Some viruses lack specificity for host cells (e.g., retroviruses).
   • Genes integrate randomly and thus may disrupt host genes (e.g., retroviruses).
Nonviral delivery systems are being actively investigated as they do not have some of the drawbacks of viral vectors. Liposomes or lipoplexes (very small spherical lipid bilayers) have been designed to hold plasmids for the transfer of genes. This transfer has been less efficient than that mediated by viruses. Amino acid polymers containing

genes to be transferred are also being investigated. Direct injection of DNA into cells without any "wrapping" (see Microinjection, Chapter 13) is also being employed to transfer genes.

Nonviral vectors have been successful in transferring genes to cells in the laboratory but as yet they have only rarely been successful in delivering them into the body of the patient.

IN VIVO GENE THERAPY

This is the delivery of a gene(s) to an organ or tissue of an individual to alleviate a genetic disorder. This approach, which is in its infancy, involves various gene delivery vehicles, including adenoviruses, retroviruses, and herpesviruses. The viruses are used to infect individuals having the genetic deficiency and is dependent on the virus being able to infect specific tissues. Various approaches to target the virus to specific tissues is the subject of much research.

ANTISENSE THERAPY

The idea of this approach is to prevent or lower the expression of a specific gene. It involves the *in vivo* (in the animal) treatment of a genetic disease by the blocking of translation of a protein with an RNA or DNA sequence that is complementary to all or a portion of a specific mRNA. Although it is still largely experimental, a number of gene therapy companies have, as a result of the high interest, acquired considerable venture capital.

GERM-LINE GENE THERAPY

This involves delivery of a gene(s) to a fertilized egg or an early embryonic cell. The transferred gene(s) is thus present in all nuclei of the cells of the mature individual. This gene(s) alters the phenotype of the individual, i.e., corrects the genetic disorder. There is little if any activity in this field at present.

Stem cells, i.e., cells originating in bone marrow, have been successfully corrected by using viruses to deliver the normal gene to cells cultivated *in vitro*. The defective cells are eliminated from the bone marrow by specific drugs and then repopulated by transplantation with the "corrected" cells having the normal gene (see *Ex Vivo Therapy*).

IN VITRO FERTILIZATION

This procedure is, strictly speaking, not gene therapy but it has been used to prevent production of individuals with genetic disorders. Sperm and egg are mixed in the laboratory and at an early stage of embryonic development (up to eight cells), one cell is removed for DNA analysis. A number of such cells from other incipient embryos can be examined for genetic effects within a short time. The embryo with the normal genetic constitution is then implanted in the mother's uterus. The resulting child should not have the genetic defects for which its embryonic cells were screened.

Chapter 16

ETHICAL ISSUES AND THE NEW BIOLOGY

The following is a brief summary of the principal ethical issues of the New Biology.

HUMAN EMBRYO RESEARCH

This type of research is permitted under certain conditions in some countries, including the United States, but is illegal in others. However, in the United States the use of federal monies for this research is prohibited. Most opposition to this research comes from religious groups that strongly oppose abortion. For most of those in opposition, the basic question is, when does the human embryo become a human being? The contention is that if it is a human being it is entitled to human rights.

GENETIC SCREENING

There is a general concern that genetic screening will lead to the practice of eugenics, i.e., the improvement of the human race by genetic means.

A number of genetic screening programs are employed for individuals ranging from the unborn to the adult. These have been described in Chapter 9. Screening of embryos/fetuses for the chromosomal abnormality leading to Down syndrome has been widely accepted. However, the advisability of screening the newborn for a battery of genetic diseases has been seriously questioned because of the distress the screening results can cause parents.

The identification of genes that can lead to nonfatal disorders such as Marfan syndrome (affected President Lincoln, see Chapter 9) can stigmatize an individual. Screening for treatable diseases such as phenylketonuria has obvious value in that controlling the individual's diet allows him/her to overcome the disorder. It is generally conceded that the genetic screening of adults should only be carried out with the consent of the individual and that the result should be kept confidential.

There are pros and cons of knowing the likelihood of inheriting a genetic disorder. Among these are possible social discrimination as well as implications for insurance eligibility and employment.

RELEASE OF GENETICALLY ENGINEERED MICROBES, PLANTS, AND ANIMALS INTO THE ENVIRONMENT

There is a wide concern for the possibility of damaging the environment by upsetting the ecological balance as a result of the release of genetically-engineered microorganisms (GEMS). There is the view that the only transformed microbes that should be released are those that cannot survive in the environment without human support. Unfortunately we don't know enough about the survivability of some of the many organisms that have been transformed. There are strict federal guidelines governing the use/release of genetically-engineered microbes.

Another serious concern is that the release of so many transformed microorganisms may eventually result in the emergence of some as human or animal pathogens. Studies to date on transformed microorganisms funded by the Environmental Protection Agency (EPA) suggests that their extrachromosomal material puts them at a disadvantage relative to normal microorganisms.

MODIFICATION OF GENES IN THE HUMAN GERM-LINE

By altering genes in the human germ-line or by adding genes, it may be possible in the future to transform human cells. For some people, such possible modifications raise again the specter of the practice of eugenics.

CLONING OF HUMAN BEINGS

That it may be possible to clone human beings was indicated by the recent successful cloning of an adult sheep (see Chapter 13). The thought of cloning humans has widespread fundamental opposition, although there are some who think that it is inevitable and could have some benefits. The United States has prohibited the use of federal funds for research on human cloning, and a number of countries have laws prohibiting such practice.

There is less objection to the cloning of animals. Some of the benefits of cloning adult mammals are mentioned in Chapter 13.

Given the almost instinctive objection to creating genetically identical human beings, it would seem that it will be many years before the practice will gain appreciable acceptance. In view of the very controversial nature and importance of the subject, it would seem incumbent upon governments to provide guidelines and laws relating to both embryo research and human cloning.

Chapter 17

IMPORTANT MILESTONES IN GENETICS AND RELATED FIELDS*

Myriad research studies from a number of disciplines over more than a century have contributed to the spectacular progress of genetics and related areas. Listed below are some of the discoveries deemed to be particularly significant.

1865 The studies of **Gregor Mendel**, which went undiscovered for 30 years, laid the basis for modern genetics. His observations resulted in a set of laws that explain the inheritance of biological characteristics. The basic idea of these rules is that each heritable trait of an organism is determined by a discrete unit in the cell called a gene.

1871 F. Miescher discovered DNA in the nuclei of certain pus cells.

1900 Reports of Mendel's studies were discovered.

1903 W. Sutton proposed that genes reside on chromosomes and that they carry hereditary information.

1910 T.H. Morgan working with *Drosophila* showed that a gene could change by mutation. His studies lead to procedures for gene mapping.

1911 E. von Dungem and **L. Hirshfeld** demonstrated the inheritance of A, B, and O blood types.

---

* The authors gratefully acknowledge the help of *A Dictionary of Genetics* (Fifth Edition, by R.C. King and W.D. Stanfield, Oxford University Press, New York, 1997) in the preparation of this chapter.

1927 H. Muller and **L.J. Stadler** independently demonstrated that X-rays could cause mutations in *Drosophila*.

1928 F. Griffith working with smooth and rough variants of the bacterial pathogen *Pneumococcus*, demonstrated a substance called transforming principle that led to the conversion of rough pneumococcal strains to smooth strains.

1941 G.W. Beadle and **A.L. Tatum** working with the mold *Neurospora crassa* proposed that one gene was responsible for one enzyme.

1943 S.E. Luria and **M. Delbruck** demonstrated that bacteria undergo spontaneous mutation.

1944 O.T. Avery, C.M. Macleod, and **M. McCarty** characterized the transforming principle of the pneumococcus as essentially DNA and concluded that the hereditary material was not protein.

1945 S.E. Luria demonstrated that mutations occur in bacteriophages (bacterial viruses).

1949 L. Pauling found that the genetic disorder sickle-cell anemia was due to an abnormal hemoglobin molecule.

1950 E. Chargaff demonstrated that for DNA, the numbers of adenine and thymine bases are always equal as are the numbers of guanine and cytosine bases. **H.F. Wilkins** and **R.F. Franklin** demonstrated with x-rays that DNA had a regular structure.

1952 A.D. Hershey and **M. Chase** demonstrated that the DNA of bacteriophage enters the bacterium and the protein does not. It followed that since the bacteriophage reproduces the genetic material was DNA.

1953 J.D. Watson and **F. Crick**, based on the x-ray photographs of Wilkins and Franklin, proposed that the structure of DNA was comprised of two helically-intertwined chains linked together by hydrogen bonds between the purines and pyrimidine bases.

1954 E.L. Wolman and **F. Jacob** described the "plasmid" (Lederberg, 1952) nature of the F (fertility) factor responsible for transmitting genes from a donor bacterium to a recipient bacterium.

1956 J.H. Tjio and **A. Levan** showed that humans possessed 46 diploid chromosomes.

1959 J. Lejeune, M. Gautier, and **R. Turpin** showed that Down syndrome was due to an extra chromosome 21 (trisomy 21).

1961 L. Orgel, S.S. Brenner, and **F. Crick** working with bacterial viruses demonstrated that the genetic code was triplet.

1970 S. Mitzutani and **H. Temin** showed that the central dogma of Crick had to be modified because of the enzyme reverse transcriptase which is able to synthesize DNA from RNA. **H. Smith** isolated the first restriction endonuclease.

1972 S.N. Cohen, A.C.Y. Chang, and **L.H. Hsu** demonstrated that *E. coli* could take up plasmids engineered with an antibiotic resistance gene and that the resulting bacterial population could be identified by virtue of their antibiotic resistance.

1973 S. Cohen, A.C.Y. Chang, H.W. Boyer, and **R.B. Helling** prepared the first recombinant DNA molecule.

1977 F. Sanger, and **M. Maxam,** and **W. Gilbert** independently, devised methods for sequencing DNA.

1983 K.B. Mullis described the polymerase chain reaction.

1985 A.J. Jefferies, V. Wilson, and **S.L. Thien** employed DNA fingerprinting (Variable Number of Tandem Repeats) which resulted in the conviction of a murderer.

1989 L.-C. Hsui, with more than a score of collaborators, identified the cystic fibrosis gene.

1990 W.E. Anderson and collaborators were the first to use gene therapy on a patient with adenosine deaminase deficiency.

1991 The Human Genome Project was officially initiated.

1993 After 10 years of intensive research, **M.E. MacDonald** and an international group of 56 colleagues identified the Huntington disease gene.

1994 Y. Miki and 44 associates identified the gene *BRCA1* that when mutated predisposes to breast and ovarian cancer.

1995 R. Sherrington, P.H. St. George-Hyslop and 36 colleagues isolated and identified a gene on chromosome 14 which is considered to be responsible for as much as 80 percent of early-onset, familial Alzheimer's disease.

1996 C. Bult and 39 associates showed that most of the genes making-up the genome of the archeon *Methanococcus jannaschii* are distinct from those of the genomes of both prokarotes and eukaryotes.

1997 I. Wilmut and associates demonstrated that the cloning of a sheep (Dolly) was possible by replacing the nucleus of an unfertilized egg with the nucleus from a mature sheep. The egg was "fertilized" and the resulting embryo implanted in a surrogate ewe.

GLOSSARY

allele An alternative form of a gene situated at the corresponding site on a homologous chromosome. Allelic genes separate during sexual reproduction and enter one gamete (sex cell) or another. They are responsible for alternative hereditary characteristics, e.g., hair color; the allele for dark hair may be from the male and the allele for light hair from the female.

amino acids Each amino acid consists of an amino group (NH2), an organic acid group (-COOH), and a unique side chain. There are 20 different amino acids commonly found in all living things. Proteins consist of one or more polypeptide chains. A polypeptide chain is made up of amino acids. The 20 amino acids found in proteins are: Alanine (Ala), Arginine (Arg), Asparagine (Asn), Aspartic acid (Asp), Cysteine (Cys), Glutamic acid (Glu), Glutamine (Gln), Glycine (Gly), Histidine (His), Isoleucine (Ile), Leucine (Leu), Lysine (Lys), Methionine (Met), Phenylalanine (Phe), Proline (Pro), Serine (Ser), Threonine (Thr), Tryptophan (Trp), Tyrosine (Tyr), Valine (Val).

amplification Increasing the number of copies of a DNA sequence.

analogue A compound that is slightly different structurally from a biological molecule such as an amino acid, a purine or a pyrimidine.

antibody A protein, one of five classes of immunoglobulins, produced by special blood cells. Antibody has the capacity to recognize and bind to foreign molecules such as those on the surface of pathogenic microorganisms.

antigen A substance, usually external to the body but occasionally within the body, which the immune system recognizes as foreign or not-self. When thus recognized it elicits a specific antibody which eventually interacts with it.

ATP (Adenosine triphosphate) An important energy-rich molecule occurring in all cells. It serves as a source of energy for physiologic reactions. ATP less its energy = ADP (adenosine diphosphate).

autoradiograph A picture produced by autoradiography.

autoradiography A procedure in which a photographic film is used to detect ionizing radiation from a radioactive label.

autosome A chromosome other than a sex chromosome.

auxotroph A mutant that differs from the wild strain (prototroph) in having an additional nutritional requirement.

avidin A protein with a high affinity for biotin. It can be labeled with a number of marker materials including alkaline phosphatase, horseradish peroxidase, fluorescent dyes, ferritin, and radioactively labeled molecules.

bacteriophage (phage) This is a DNA or RNA virus that infects bacteria. Bacteriophages are used as cloning vectors. Some have genomes with as few as several thousand base pairs. The genomes of phages (e.g., M13 and lambda phages) are manipulated by replacing or inserting DNA. Lambda phage can insert its genome into the host's and be replicated along with it. Its genetic material has been sequenced and consists of 48,514 bases.

base That portion of the a DNA or RNA nucleotide that is distinctive. The bases are: adenine, thymine, guanine, cytosine, and uracil.

base pair (bp) Two complementary bases, one in each strand of a double-stranded nucleic acid molecule. They are attracted to each other by weak chemical bonds. They are paired as follows:
   •DNA: Adenine with Thymine, and Guanine with Cytosine
   •RNA: Adenine with Uracil, and Guanine with Cytosine
The chromosome of *E. coli* has approximately four million base pairs and is about lmm long.

biotin A small vitamin with two binding sites, one of which can bind covalently with a nucleic acid leaving the other to form a strong bond with the protein avidin which in turn can be bound to enzymes. Biotin is used as a label for nucleic acid probe detection and also in enzyme immunoassays.

blastocyst stage The stage of mammalian development when implantation into the uterine wall takes place.

blunt end (flush end) The end of a double-stranded DNA molecule in which strands do not extend beyond one another.

blunt-end cut Phosphodiester bonds in the backbone of duplex DNA are cleaved between the corresponding nucleotide pairs on opposite strands so that no nucleotide extensions occur on either strand.

carrier In genetics it refers to an individual with one mutant allele and one corresponding normal allele. The individual's phenotype is normal.

catabolite A small molecule resulting from the breakdown of a larger molecule in a living organism.

cDNA A duplex DNA complement of a messenger RNA sequence which has been synthesized *in vitro* by reverse transcriptase and DNA polymerase.

cellular immunity (cell-mediated immunity) A form of immunity medicated by T lymphocytes. It is particularly important in defending against intracellular microbes such as *Salmonella*.

centromeres Characteristic dense staining regions of chromosomes that contain no genes. They are useful as genetic landmarks.

chimera At the molecular level, a recombinant DNA molecule composed of DNA sequences from more than one organism.

chromatin The material within the nucleus of which chromosomes are composed. It consists of proteins (mostly histones), DNA, and small amounts of RNA.

cleave (cut, digest) To break the phosphodiester bonds of double-stranded DNA. It is usually done with a type II restriction endonuclease.

clone/cloning A group of identical cells descended from a single cell. A population of cells that all carry a cloning vehicle with the same insert DNA molecule, e.g., the gene for the *Pasteurella* dermonecrotoxin has been cloned in *E. coli.* The verb refers to the process of creating a group of identical cells, generally those containing identical recombinant DNA molecules.

cloning site The location of specific DNA sequences in a cloning vector where DNA can be inserted.

cloning vector The DNA of a small plasmid, bacteriophage, cosmid, animal virus, etc., used to transfer a DNA fragment from the "test tube" into a living cell.

codon Three nucleotides whose order corresponds to 1 of 20 amino acids. Some special codons do not code for any amino acids but act as stop signals. In some instances different codons may code for the same amino acid and are termed redundant.

complementary For example, in the double helix the sequence of bases A C T in one DNA strand is complementary to the sequence T G A in the other.

congenital A congenital condition is one that is present at birth.

copy number The average number of a specific kind of plasmid molecule in a cell. The copy number may range from one to as many as 300.

cosmids They are hybrids between a phage DNA molecule and a bacterial plasmid. A plasmid that carries a cos site can accommodate a fairly large amount of DNA. Colonies are formed rather than bacteriophage plaques.

covalent bonds Some atoms can share electrons with other atoms to form covalent bonds, e.g., hydrogen (1 bond), carbon (4 bonds). Carbon forms covalent bonds with most carbon compounds.

crossing-over (recombination) An interaction between two homologous chromosomes whereby portions of the chromosomes (double-stranded DNA) are exchanged.

Dalton A unit of mass used to express masses of atoms, molecules and nuclear particles. It is equal to one-twelfth of the weight of the carbon 12 atom; it is also called atomic mass unit.

denature To denature is to alter the natural state of proteins or nucleic acids. Denaturation unfolds proteins, and unwinds and separates the two strands of the double helix of DNA and RNA.

deoxyribose A five-carbon sugar (pentose) found in DNA and having one oxygen atom less than ribose.

dicotyledons One of two classes of flowering plants distinguished by having two seed leaves (cotyledons) within the seed. Examples are potatoes, peas, beans, roses, hardwood trees, etc.

dideoxynucleotide (ddNTP) A nucleoside triphosphate without hydroxyl groups on both 2' and 3' carbons of the pentose sugar. It thus prevents further chain elongation when incorporated into a growing polynucleotide (see Chapter 8).

dimer A chemical entity made up by the association of two monomeric subunits, e.g., two polypeptide chains in a functional enzyme.

diploid Diploid cells have pairs of homologous chromosomes.

DNA construct A DNA molecule inserted into a cloning vector, usually a plasmid.

DNA hybridization or **homology** (see Chapter 4).

DNA insert A segment of DNA that is joined to a recombinant DNA vector with the purpose of cloning.

DNA ligase An enzyme that joins two DNA molecules end-to-end. It is used in ligating DNA fragments into a vector.

DNA polymerase An enzyme that polymerizes DNA. It makes new DNA using the information from old DNA. Taq polymerase is a DNA polymerase used in PCR (see Chapter 6).

dominant gene An allele that expresses its phenotype in the presence of its homologous recessive allele.

downstream The stretch of nucleotides of DNA that lie in the 3' direction of the site of initiation of transcription, which is designated as +1. Downstream nucleotides are marked with plus signs, e.g., +2, +10 Downsteam position refers to nucleic acid sequences on the 3' side of a given site on the RNA or DNA molecule.

duplex DNA Double-stranded DNA.

electrophoresis (see **gel electrophoresis** and **pulsed field gel electrophoresis**) Separation of molecules by their net electrical charge.

electroporation A method for increasing DNA uptake by cells by prior exposure to high voltage which results in the temporary formation of small pores in the cell membrane.

ELISA (Enzyme-linked immunosorbent assay), Enzyme immunoassay A serological procedure for the detection and measurement of antibody or antigen tagged with an enzyme. The presence or absence of the antibody or antigen, depending on which is being tested for, is determined by the addition of the appropriate substrate. The presence of the enzyme with its action on the substrate results in a detectable color. The assay is remarkably sensitive.

end-filling This involves changing a sticky end to a blunt end by enzyme synthesis of the complement to a single-stranded extension. It is used in the labeling of DNA molecules that have sticky ends, e.g., in autoradiography.

endoplasmic reticulum (ER) A membrane network in eukaryotic cells which forms a link with the plasma membrane and the nuclear membrane. If the outer surface is studded with ribosomes, it is said to be rough-surfaced.

enteropathogenic Capacity to cause intestinal disease.

enterotoxin A toxin that affects the intestine, e.g., causing diarrhea.

enzyme A protein that catalyzes a chemical reaction. It is not permanently altered by the reaction and can be used repeatedly.

epitope (see antigenic determinant)

ethidium bromide This compound is used to stain gels to make DNA visible. Bands showing the position of the different sizes of DNA fragments are visible (i.e., they fluoresce) under ultraviolet light.

eugenics The idea that has as its objective the improvement of humans by genetic means.

eukaryotic cells The cells of animals, plants, and fungi. They have a well-defined nucleus which contain the chromosomes.

exon The mRNA sequence remaining after excision (removal of introns) has taken place. The coding sequence of the genome that is not deleted by the splicing of the primary RNA transcript, but appears in mRNA and is translated into protein.

ferritin It is a protein of 700 kDa containing 23 percent iron as ferric phosphate or ferric hydroxide. Ferritin can be linked to immunoglobulin which can attach to antigens on cell surfaces. The high iron content makes visualization with the electron microscope possible.

flow cytometry In this procedure cells are passed through a laser beam. When the cells pass through a number of measurements are made by detectors while light is being scattered in various directions. A number of properties are measured by a multi-parameter analysis of signals and a three dimensional picture is obtained. Fluorescent dyes can be used. Various cell populations can be differentiated including chromosomes. Bacteria can be detected in urine and blood.

fluorescent dyes These are used to tag or label compounds. They are visible under ultraviolet or other light sources, e.g., laser.

free radicals They are unstable, highly reactive molecules that attack a variety of organic structures including DNA. They are strong oxidizing agents capable of disrupting the structure, resulting in loss of function.

gel electrophoresis (See Figure 5.1) A method for separating molecules, proteins and nucleic acids, based on their electric charge and size. The molecules or portions of them are pulled through a gel (e.g., agarose and/or polyacrylamide) by means of an electric current. The speed with which they move through the gel depends upon size and electric charge; the smaller the size the faster they move.

The size of DNA fragments is obtained by including in the run standard fragments of DNA, the sizes of which are known.

Cathode = negative, anode = positive; movement of negatively charged DNA is towards the positive electrode.

The bands are made visible by staining with ethidium bromide (a fluorescent compound) and then observing under ultraviolet light; the stained bands fluoresce.

gene library (Gene bank, Genomic library) (see Chapter 4.) A genomic library or bank is a set of recombinant clones that contain all or most of the genes present in an individual organism.

genetic marker An identifiable fragment of DNA that is located in a definite region of a chromosome. It serves as a signpost in genetic mapping. RFLPs and VNTRs are important examples.

genome The complete set of genes of an organism.

genotype The genetic constitution or sum of the genes of an organism as opposed to its overt character make-up or phenotype. It is also defined as the combination of alleles present at a locus, or several loci.

haploid cell One that contains only half or one set of the usual diploid number of chromosomes.

heterologous probing This involves the use of a labeled nucleic acid molecule to identify related molecules using hybridization probing (see Chapter 4).

heterozygous A state in which two different versions of a gene (allele) occur in a diploid genome. Heterozygous organisms are often called heterozygotes.

histones Proteins of major importance in the packaging of eukaryotic DNA in chromosomes.

homologous chromosomes Chromosomes that pair with one another; one from the female, the other from the male. Homologous chromosomes have the same pattern of genes along the chromosome although the genes may differ.

homozygous The same version of a particular gene (allele) occurs on both chromosomes of a diploid genome. Homozygous organisms are often referred to as homozygotes.

humoral immunity The form of immunity mediated by antibodies found in body fluids such as blood serum.

hydrogen bonds The weak electrostatic attraction between molecules that have hydrogen atoms bound to electronegative atoms such as: F, N and O.

interferon Several closely related proteins produced principally by cells of the immune system and which induce resistance to viral infections.

intron A noncoding section of a gene removed from RNA before translation. The mRNA of bacteria do not contain introns.

IPTG Isopropyl-ß-D-thiogalactoside an inducer of the *lac* (lactose) operon. It is frequently used to induce cloned genes that are under control of the *lac* repressor-*lac* promoter system.

isotopes Atoms of a chemical element having the same atomic number but different atomic weights. They may be stable or unstable (radioactive).

jumping gene (see **transposon**)

"junk" DNA "junk" DNA is nonfunctioning genetic material that makes up about 95 percent of the human genome. It serves no coding purpose but is possibly protecting expressing genes from deleterious mutations.

karyotype The number, size and shape of the set of chromosomes of a cell.

killer T-cell A T cell (lymphocyte derived from the thymus) that specifically targets and destroys certain infected cells and cancer cells.

kilobase pair (kb) A kilobase pair is a segment of nucleic acid containing one thousand base pairs.

label A compound or atom that is either attached to or incorporated into a macromolecule and is used to detect the presence of a compound, substance, or macromolecule in a sample. Examples:
- a radioactive nucleotide in a nucleic acid molecule
- biotin labeled nucleotide (biotinylated probe); detected by avidin coupled to a fluorescent dye
- horseradish peroxidase complexed to DNA and used as a probe
- alkaline phosphatase, acridium ester, and others
- fluorescent dyes tagged to various molecules.

Nick translation and end-filling are methods used in labeling.

lambda (see bacteriophage)

ligase (see DNA ligase)

locus A location on a chromosome (or chromosome pair) occupied by a gene or by two alleles of the same gene.

M13 (see **bacteriophage**) A filamentous bacteriophage that infects *E. coli* and is used as a cloning vector.

marker An identifiable physical location on a genome, such as a gene or a restriction fragment length polymorphism whose inheritance or origin can be traced. The ampicillin resistance gene is a selectable marker.

meiosis Cell division in which the number of chromosomes is reduced by half and as a result gametes (sperm and ovum) have half the number of chromosomes as do somatic cells.

methylation The addition of a methyl group(s) to a macromolecule. Methylation of DNA can render one or more genes within a chromosome silent.

monoclonal antibody A single type of antibody, produced by a hybridoma cell line, and directed against an antigenic determinant (epitope). The hybridoma cell line was formed by the fusion of a lymphocyte cell and a myeloma cell (from a plasma cell cancer).

mutagen A physical or chemical agent that alters the nucleotides of the DNA of an organism.

mutation Any change that alters the sequence of nucleotide bases of the DNA of an organism.

mycoplasmas A group of very small bacteria characterized by the absence of a cell wall. There are both free-living and pathogenic species.

nick A break in the phosphodiester backbone in a single-strand of a molecule of double-stranded DNA. Labeled nucleotides can be introduced into the nick.

nucleotide It consists of:
- a base (a purine or pyrimidine) which lies flat like steps of a staircase
- a suger   )
) the backbone
- a phosphate)

DNA has four different nucleotides: adenine(A), thymine(T), guanine(G) and cytosine(C); adenine and guanine are purines, thymine and cytosine are pyrimidines; uracil of RNA is a pyrimidine. The sugar in DNA is deoxyribose; in RNA it is ribose.

oligonucleotide A short piece of RNA or DNA derived from an organism or synthesized chemically. It contains three or more nucleotides; those used in gene cloning are usually less than 100 nucleotides long.

oncogene A gene that plays a role in the cell division cycle. Often mutant forms of oncogenes can cause a cell to grow in an uncontrolled manner which may lead to cancer.

operator A specific region of the DNA molecule capable of interacting with a repressor and thus controlling the functioning of an adjacent gene.

organelle A microscopic structure within a plant or animal cell that has a particular function.

palindromes Sequences that read the same in both directions. Most recognition sites of restriction endonucleases are palindromes, e.g., the recognition sequence of *EcoRI* (*E. coli*) is:

5' G A A T T C 3'
3' C T T A A G 5'

pathogenic Having the capacity to cause disease.

PCR (polymerase chain reaction) (see Chapter 6).

peptide A compound consisting of two or several amino acids.

phenotype The sum of observable properties of an organism. The pheno-type results from the interaction of the genotype and its environment.

phosphorus 32 A radioactive isotope of phosphorus.

physical map A map of the locations of landmarks on DNA. It includes restriction enzyme cutting sites, RFLP markers, genes, etc. The lowest-resolution physical map is the banding patterns of the chromosomes. The highest-resolution map is the complete nucleotide sequence of the chromosomes.

plasma membrane The outer lining membrane of the eukaryotic cell.

plasmid amplification It is often difficult to isolate plasmid DNA because it is a small part of the total DNA of the cell. The aim of plasmid amplification is to increase the copy number of a plasmid. Plasmids with a copy number of more than 20 can replicate in the absence of protein synthesis. Chromosomes need protein synthesis to replicate. When a protein synthesis inhibitor such as chloramphenicol is added to a culture, copy numbers of several 1000 can be obtained.

plasmid fingerprinting (plasmid profile) Plasmids from various bacteria–even from within a species–vary in number and size, i.e., the number of kilobase pairs of DNA. Their size is determined by their electrophoretic migration in a gel. If the plasmid profile is questionable, plasmids can be further distinguished by restriction endonuclease (see Chapter 4) analysis. Similar digestion patterns virtually assure plasmid identity.

Plasmid fingerprinting is useful in identifying varieties within a species. This is of value in tracing the spread of bacteria in epidemiological studies.

pluripotency The property of a cell by which it can give rise to many cell phenotypes. The stem cell is an example.

polyacrylamide gel (PAGE) A gel made from polyacrylamide and used to separate macromolecules based on size and charge.

polymerase chain reaction (PCR) (see Chapter 6).

polymorphisms Variant forms of a particular gene that occurs simultaneously in a population, e.g., one polymorphism in the hemoglobin gene which results in sickle-cell anemia, persists in some populations because it confers resistance to malaria. GTC = glutamic acid = normal hemoglobin; GTG = valine = sickle-cell anemia.

polypeptide A chain of many amino acids joined together by peptide bonds. A polypeptide may have anywhere from about 50 to several thousand amino acids.

3'(Prime) end, 5'(Prime) end (see Figure 1.3) The backbone of a nucleic acid molecule consists of repeating phosphate and sugar subunits. On one side of the sugar the phosphate is linked to the 5' carbon and on the other side to the 3' carbon. When the molecule is broken, the break usually occurs between the phosphate and the sugar, thus 2 different ends are produced. If only the sugar is considered a 5' carbon will be at one end a 3' carbon at the other. Thus a double-helical region can be represented as follows:

5'-AAGCTG-3'
3'-TTCGAC-5'

primer (see Chapter 5)

prion An infectious protein that appears to be able to direct its own replication in animal cells without nucleic acids. It is considered the cause of scrapie in sheep, bovine spongiform encephalopathy in cattle, and other fatal, chronic diseases.

probe A sequence of labeled DNA or RNA used to detect the presence of complementary nucleotide sequences by hybridization. Probes are used widely in recombinant DNA technology for the identification DNA segments. In clinical microbiology they are used in the identification of a pathogen by probing its genetic constitution.

   A gene probe can be used to find ribosomal RNA which is abundant throughout the cell. A single stranded segment of DNA can be used to probe for a complementary strand of ribosomal RNA.

prokaryotic cells The cells of bacteria and blue-green cyanobacteria; the nucleus is not well defined.

promoter A short nucleotide sequence on DNA where RNA polymerase binds and initiates transcription.

pronucleus The haploid nucleus of a sperm, egg, or pollen grain.

protein A folded polypeptide chain or several folded polypeptides. The unique sequence of amino acids is called the primary structure and it determines how the polypeptide folds up into a characteristic shape: secondary, tertiary, or quaternary structure. The three-dimensional form is responsible for the biological properties and functions of proteins.

protein A A protein on the cell wall of some strains of *Staphylococcus aureus* that binds specifically to immunoglobulin G molecules.

protoplast A cell without a cell wall.

protozoa Protozoa are a category of eukaryotic microorganisms in the kingdom Protista. They comprise a diverse group of single-celled, animal-like organisms that are considerably more complex than prokaryotes (e.g., bacteria). Species include the causes of amoebic dysentery, malaria, and sleeping sickness.

pulsed field gel electrophoresis (PFGE) In this type of gel electrophoresis an electric field alternates rapidly between two pairs of electrodes; each pair is set at a 45 degree angle. It makes possible the separation of large molecules or fragments up to several 1000 kb that can't be separated by conventional gel electrophoresis. Eukaryotic chromosomes can be separated by this procedure.

purine A two-ring nitrogenous base, e.g., adenine and guanine.

pyrimidine A type of one ring (monocyclic) nitrogenous base, such as thymine, uracil, and cytosine.

radioactive isotope An isotope with an unstable nucleus that emits ionizing radiation (beta and gamma rays).

radioactivity The spontaneous disintegration of an atom, with the emission of alpha, beta, and/or gamma rays.

recessive gene An allele that is expressed in the phenotype only when two such genes are present, e.g., cystic fibrosis results when the individual inherits the two recessive genes responsible for the disease. Some other recessive genetic disorders are referred to in Chapter 9.

recombinant DNA molecule A DNA molecule containing two or more regions of different origin formed by cutting and splicing techniques.

recombination (crossing-over) The reciprocal exchange of DNA between two chromosomes or DNA molecules by a breakage and exchange process. In homologous recombination the sequences exchanged are between two very similar DNAs. In heterologous recombination the combination is between DNAs unrelated in nucleotide sequence.

replicon A genetic element that acts as an autonomous unit in DNA replication, e.g., the bacterial chromosome. Eukaryotic chromosomes in contrast contain many replicons in series.

repressor A protein encoded by a regulatory gene. It can either combine with the operator blocking RNA polymerase access to the promoter and thus repressing transcription, or combine with the inducer that permits transcription of structural genes.

restriction endonuclease (see Chapter 4).

restriction endonuclease mapping (see Chapter 7). Specific cuts are made in DNA with restriction endonucleases. The location of the cuts are measured and oriented relative to each other to form a map. This procedure is used to obtain profiles of genomic DNA and plasmid DNA.

restriction fragment length polymorphisms (RFLP) (see Chapter 5)

restriction map A map showing the positions of different restriction sites in a DNA molecule.

restriction sites Specific recognition sequences where restriction endonucleases cleave double-stranded DNA. They are usually composed of four to six bp.

retroviruses These are RNA viruses that require the enzyme reverse transcriptase to replicate. In replicating they transfer their genetic information to complementary DNA molecules upon infecting cells. HIV (human immunodeficiency virus) is a retrovirus. One of the nine genes of HIV codes for reverse transcriptase.

reverse transcriptase The enzyme used by retroviruses to synthesize DNA from an RNA template. Retroviruses contain single-stranded RNA. Molecular biologists use this enzyme to make DNA clones from mRNA.

R factor A plasmid that carries a gene coding for resistance to one or more antimicrobial agents. An R factor is composed of two components, the resistance determinant and the resistance transfer factor (RTF) which can be transferred with or without the resistance determinant. The transfer of resistance determinants to recipient cells during conjugation is controlled by RTF.

RFLP (see **Restriction Fragment Length Polymorphism** in Chapter 5)

RNA polymerases These are enzymes that produce RNA by joining ribonucleoside triphosphates. There are several different types with different functions, e.g., types I, II, and III in eukaryotic cells.

S (Svedberg unit) The S number indicates the relative rate of sedimentation during ultra high-speed centrifugation.

Sendai virus A virus that causes a important, widespread disease of mice. It is used in cell fusion studies because infected cells tend to fuse.

shuttle vector A vector that can replicate in cells of more than one organism, e.g., in *E. coli* and a yeast.

somatic cells Cells of an organism other than sex cells.

Southern blotting (Southern transfer) This technique is used to identify specific genetic sequences of DNA. The specimen of DNA is treated with restriction enzymes to yield DNA fragments; these are separated by gel electrophoresis. The gel is blotted to a nitrocellulose or nylon membrane that is then treated with a probe consisting of DNA fragments complementary to those being sought to see if hybridization has taken place. If the probe is radioactive, autoradiography is used for detection. If a non-radioactive label is used, another detection system is employed.

Southern transfer (see **Southern blotting**).

stem cells These are large cells of the bone marrow and other tissues that can self-replicate and generate a series of cells of immunological importance.

sticky end The end of a double-stranded DNA molecule that has a single-stranded extension generated following a restriction endonuclease cut.

SV40 (Simian virus 40) A monkey virus used as a cloning vehicle. It has served as a model for gene expression and DNA replication in cultured animal cells. Its genome is a double-stranded circle of about 5200 base pairs.

telomere A set of repeated short DNA sequences at the end of each chromosome.

terminator It marks the point at the end of the gene where transcription should stop.

transfection The insertion of a gene into a cell by mechanical means, e.g., by electroporation, by calcium phosphate precipitation (sedimenting the DNA onto the cell), or by microinjection (injection into a eukaryotic cell by a fine, microscopic needle).

transformation The introduction of any DNA molecule into a living cell whereby it is integrated into the cell's genome and functionally expressed. The term transformation is also used to refer to changes in animal cells which occur when a healthy cell becomes cancerous. Such a cell is said to be transformed.

transposon A highly mobile DNA sequence that moves from one chromosomal location to another. They are sometimes called "jumping genes." Tn5 is a bacterial transposon that carries genes for resistance to neomycin and kanamycin, and genetic information for insertion and excision of the transposon. An enzyme, transposase, is responsible for the movement of the transposon.

upstream The stretch of DNA base pairs that lie in the 5' direction from the site of initiation of transcription. Usually the first transcribed base is designated +1 and the upstream nucleotides are marked with minus signs, e.g., -1, etc.

VNTR (see **Variable Number Tandem Repeats** in Chapter 5).

vaccinia virus A DNA virus originally recovered from the disease cowpox and used to vaccinate against smallpox.

virulence An expression of an microorganism's capacity to cause disease, e.g., a highly virulent bacterium has a great capacity for causing disease.

virus Viruses are very small infectious agents that invade and reproduce in eukaryotic and prokaryotic cells. Their genome may be either DNA or RNA and it is surrounded by a protein coat. The DNA genomes may be linear or circular helices. Some are single-stranded. Viruses cannot reproduce on their own. They co-opt the host cell's reproductive machinery for transcription, translation, and replication. New viruses are released from the cell which may be destroyed in the process.

X-linked recessive mutation (see **Sex-linked genetic disorders**, Chapter 9)

YAC (yeast artificial chromosome) A cloning vector system consisting of a yeast with an artificial chromosome. It makes possible the cloning of DNA fragments that can be hundreds of kilobases long. It is used in the mapping of the genome of various species.

FURTHER READING

Brown, T.A.: *Gene Cloning: An Introduction*. 2nd ed., New York, Chapman and Hall, 1990.

Drlica, K.: *Double-Edged Sword.* New York, Addison-Wesley Publishing Company, 1994.

Glick, B.R., and Pasternak, J.J.: *Molecular Biotechnology.* Washington, ASM Press, 1994.

Hubbard, R., and Wald, E.: *Exploding the Gene Myth.* Boston, Beacon Press, 1997.

Jones, S.: *The Language of Genes.* New York, Anchor Books, Doubleday, 1994.

Lee, T.F.: *Gene Future.* New York, Plenum Press, 1993.

Levine, J., and Suzuki, D.: *The Secret of Life.* Boston, WGBH Educational Foundation, 1993.

Lewontin, R.C.: *Biology as Ideology.* New York, Harper Collins, 1993.

Suzuki, D., and Knudtson, P.: *Genethics.* Cambridge, Harvard University Press, 1990.

Wills, C.: *Exons, Introns, and Talking Genes.* New York, Basic Books, 1991.

Wingerson, L.: *Mapping Our Genes.* New York, Dutton, 1990.

INTERNET RESOURCES

Readers who are interested can find a great deal of information relating to Genetic Engineering, Genetics, and Biotechnology on the Internet. Some of the popular web search engines are *Yahoo, Lycos, Archie* (for combing *FTP* sites for files and documents), *Gopher* (a file finder), and *Veronica* (used to search *Gopher* by keywords).

INDEX